Authentic Involvement

Dan Petersen

National Safety Council®

NSC PRESS

Project Editor: Patricia M. Dewey
Cover Design: Ryan Anderson, National Safety Council
Interior Design and Composition: Elizabeth MacKinney

NATIONAL SAFETY COUNCIL MISSION STATEMENT

The mission of the National Safety Council is to educate and influence society to adopt safety, health, and environmental policies, practices, and procedures that prevent and mitigate human suffering and economic losses arising from preventable causes.

DISCLAIMER

Although the information and recommendations contained in this publication have been compiled from sources believed to be reliable, the National Safety Council makes no guarantee as to, and assumes no responsibility for, the correctness, sufficiency, or completeness of such information or recommendations. Other or additional safety measures may be required under particular circumstances

06 05 04 03 02 5 4 3 2 1

Library of Congress Cataloging-in-Publication Data

Petersen, Dan.
Authentic involvement / Daniel Petersen.
p. cm
Includes bibliographical references and index.
ISBN 0-87912-232-3
1. Industrial safety—United States—Psychological aspects. 2. Accidents—United States—Prevention. 3. Behavior modification. I. National Safety Council. II. Title.
T55.3.B43 P45 2001
658.3'82—dc21 2001000797
1M0501 Product Number: 12120-0000

Contents

Preface

Real, authentic involvement is worker participation where the worker engages in activities that satisfy that person's human needs, and thus are motivational. Every person's needs are different—but most people respond to being treated like a mature, adult human being; being treated as an equal, not as a subordinate; being able to use their innate intelligence and skills daily, even hourly; being able to achieve; being given responsibility; and being recognized for doing a good job.

This definition broadens the idea of worker involvement beyond the traditional concepts. In my role as an advisor to industry in the areas of safety and organizational behavior, I have probably seen all structures under the label of involvement. Some have to do with the boss getting rid of the grunt work that he or she doesn't like to do. Some have to do with new forms of punishment and discipline (e.g., the "screw up" gives the next safety meeting).

I don't think that is what the researchers and theorists discussed in chapter two of this book had in mind when they talked about true employee involvement. We in safety in some ways have been the true leaders in employee involvement through the safety committees we have long advocated. We have also been the true laggards—by resisting the concepts of involvement in the Deming philosophies and by not using statistical process control concepts, total quality improvement concepts, or other management techniques that include employee involvement. We need to lead the way again.

Our hope in this book is simply to urge that the management process include true employee involvement because that will save lives and reduce pain from occupational injuries.

Introduction

Since the Occupational Safety and Health Administration (OSHA) was established, many companies have put the entire emphasis on the physical aspects of safety. Most managements know about OSHA and would rather not be fined. Most safety directors are (or will be) judged by the number of violations and the size of fines. Under these circumstances, it does not seem logical to place emphasis elsewhere.

As the list of influences in Figure I-1 indicates, our safety programs today are a composite of what the safety director knows, what the boss likes, what the company likes, and what the federal government wants. The needs of the company and the employees are notably absent from the list.

What does a company need? What do employees want in a safety program? These are tough questions to answer; so many safety directors choose simply not to try. It is easier to do what they themselves know, what the boss likes, what fits the company, or what Uncle Sam says. Does this approach achieve the best possible program? Probably not.

Of the four influences on safety program content, the fourth (what the federal government wants) by far predominates today. Each year since OSHA came into existence in 1970, federal-level regulations have increased (not to mention the influence of state and local levels).

Although some of these new regulations may not apply to every organization, many will, and some will be very costly to the organization. To comply with the

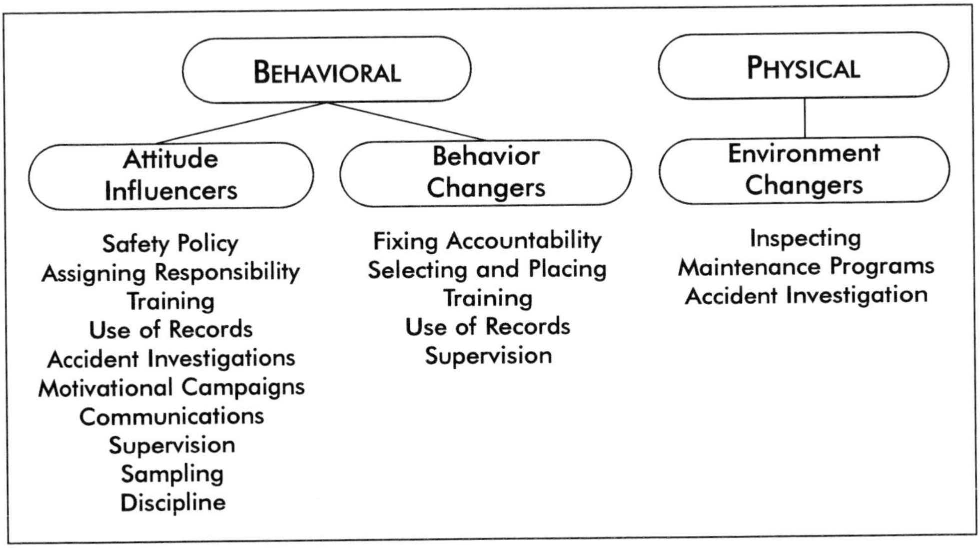

Figure I-1. Common controls in safety programs.

provisions of the process safety standard, the Americans with Disabilities Act, the hazard communication standard, and so on, is not only expensive; it is likely that these costs will never be offset by reduced losses. Actually, these pieces of legislation and others may result in increased losses because managerial time and corporate monies will be redirected away from safety toward compliance.

Worse yet, as management, including safety staff, are busy trying to put together all the paper work and documentation to comply, the real things that get results in safety are allowed to drop through the cracks.

What are the things we need to do to get results in safety? Our research is clear on this. Hundreds of benchmarking studies agree: world-class results in safety occur when an organization meets certain well-defined criteria. We know there is no one right way to achieve safety in an organization; however, for a safety system to be effective, it must meet the following criteria:

- supervisory performance and accountability
- middle management involvement and accountability
- visible top management commitment
- employee participation
- flexiblity
- a positive perception

The first three criteria focus on the management of the safety process and are achieved through a management system of holding managers at all levels accountable for specific, defined performances. In my books over the years, I have pushed the concept of accountability strongly.

The last three criteria speak to something quite different: the involvement of all employees.

In effect, the criteria for safety excellence that come from numerous research studies and numerous benchmarking studies can be boiled down to the following concepts:

- You must have tight accountability that forces management action, top to bottom.
- You must have the active involvement of as many employees as you can possibly get.

This book is about employee involvement. Once the management system is in place and working, then the key to safety excellence is the amount and kind of involvement you get.

In Part I of this book, we look at the progress in safety from the beginning; at the early nonsafety kinds of employee participation efforts; at the rationale behind the participation concepts; and then at a definition of authentic involvement—that is, involvement in real, meaningful things that help an organization to achieve excellence.

You can get great results by involving people within a traditional management structure. You can also get great results by changing your management structure. Both are discussed in Parts II and III. Part II looks at the various levels of involvement and concepts of employee empowerment: limited empowerment, shared decision making, and just getting input from employees. Part III looks at the options available by changing the structure of the organization and at some success stories.

Part IV looks at the concept of culture and describes the kinds of culture that ensure or inhibit success. One chapter is devoted to the concept of behavior-based safety, so popular in safety today.

Finally, we've included three Appendices, one describing safety sampling, a way to measure whether or not we are succeeding; one on measurement in safety generally; and one on how to assess culture, a crucial assessment before you even consider involvement.

This book, then, is about the second criterion for safety excellence—getting each employee to be a part of solving safety problems.

In writing this book I am indebted to many people, some of whom I have quoted in the book, some extensively: Scott Geller, one of the best of the behavioral gurus; Gene Earnest, for his description of the Proctor and Gamble systems; the people at P&G for their sharing over the years (Gordon Peters, Bill Blass, Rick Raymer, and others). I'm indebted to Katy Crane of Core Media; as we put together three video series over the years (mostly on management and accountabilty), we could never escape the involvement concepts and discussed them constantly.

And a special thanks to Sue Olson, Ph.D., who helped all the way on this and continued to make it better.

—Dan Petersen

I

INVOLVEMENT

1 Involvement in Safety— The History

Throughout the history of safety, we have seen many different thrusts, some of which have also been heralded as the future of safety. Some of these include the following:

- physical condition approach from 1911 to the present
- industrial hygiene approach from 1931 to the present
- unsafe act approach from 1931 to the present
- management approach from the 1950s to now
- noise control approach from 1954 to now
- audit approach from the 1950s to the present
- system safety approach from the 1960s to now
- safety management approach from the 1960s to now
- OSHA physical condition approach from 1971 to now
- OSHA industrial hygiene approach, when the OSHA chief was an industrial hygienist
- other OSHA approaches, depending on the year, and the current emphasis
- ergonomics approach, anticipating the OSHA standard
- safety program approach, anticipating the OSHA standard
- environmental approach
- total quality management (TQM) and statistical process control (SPC) thrust
- and now a behavioral approach

None of these were fads. They were real attempts to control losses. They are all still with us today as a part of our technology. They are layers of things that must be done. Now since safety staffs are much smaller than they used to be, management staff has been cut in half, and the employee staff has been downsized, the question is which layer should we work on today? Is the behavioral-based layer the one, and if so, which layer should we ignore?

This is what has happened within safety; but to a large degree we and our efforts are governed by external forces—things that are going on outside the profession, as in the external environment. The following examples illustrate the changes:

- Legislative changes dictate what we must do; new laws at the federal, state, and local levels are being promulgated almost daily.
- Enforcement of the laws that do exist has changed; for instance, guidelines have been issued, with enforcement under the General Duty Clause.
- There is a criminal liability potential at the federal, state, and local levels.
- The protections afforded by workers' compensation legislation have broken down, allowing lawsuits against companies and individuals.
- The definition of what is considered to be an accident has broadened under workers' compensation.

We are also largely influenced by internal changes in organizations. Companies are managed much differently today than they were fifty, forty, or even ten years ago, as illustrated by the following changes:

- Computers have changed almost everything, often putting more clerical work on managers than before.
- Theory X and Y concepts have changed how we manage people.
- Management by objectives (MBO) concepts have shifted decision making to lower levels in organizations.
- Management styles have changed from the old top-down concepts.
- Situational leadership has changed our whole concept of how to manage (the One Minute Manager concepts, for instance, of the 1970s).
- Theory Z and Japanese management theories of the 1980s introduced more change.
- Others include total quality management (TQM) concepts, statistical process control (SPC) concepts, Deming philosophies, reegineering, self-directed work teams, and employee ownership.

WHAT HAS BEEN LEARNED?

We've changed what we do because the external environment is totally different and because the internal environment has gone through as least three major turnarounds: from classical management to human relations management to situational management to cultural assessment. Also, safety management has learned some things during these years, including the following:

- About accident/incident causation, theories have moved from the old domino theory that stated accidents are caused by unsafe acts and/or unsafe conditions, to a newer theory that suggests that accidents are caused by a combination of management system failure and human error. Human error is also often caused by a management-created environment that rewards risk taking and may even set people up for accidents through downsizing, more work with fewer people, forced overtime, etc.
- About accident/incident control, safety management has moved from the standard safety program of the past to a recognition of the fact that there is no one right way, no magic bullet, and reams of research and benchmarking studies prove it. These studies clearly demonstrate the following points:
  - Management is the key.
  - There is no one right way.
  - The culture dictates what works.
  - Criteria for safety excellence underlay world-class organizations.

But a major discrepancy exists between what we know and what we do. We know that management is the key, but in behavioral approaches we make the employee the key, allowing the probability of management abdication. We know that there is "No One Right Way," but OSHA wants a standard safety program that fits all. We know that the culture dictates success, but usually we don't even assess the culture of an organization. We know that our accident statistics are almost always invalid, but we use them as a basic measurement of safety program success. We know that there is no magic pill out there, but we continue to buy these magic pills and silver bullets, particularly if they cost enough.

THE BEHAVIORAL APPROACH

In 1970, Frank Bird and Dr. Larry Schlesinger, in an article in the *Journal of the ASSE,* introduced the concept of safe behavior reinforcement to the profession of safety. Bird and Schlesinger did not invent the concepts but borrowed well-known concepts from the field of psychology and suggested that perhaps we might be able to apply them.

Actually, behavioral concepts had been introduced even earlier. The behavioral concepts that are involved stem from the works of John Watson, a psychologist who wrote about behaviorism as early as 1910; from Ivan Pavlov, who experimented with classical conditioning in the 1920s; and from B.F. Skinner's operant conditioning concepts of the early 1940s.

Watson insisted that psychologists should concentrate on observable behavior; thus he provided the foundation for behavior-based safety. Pavlov experimented with paring two stimuli (such as meat and a bell for dogs) showing how, over a period of time, behavior can be influenced by simple words. Skinner showed that behavior can be changed (thus the term behavior modification) by attaching a consequence immediately following the behavior (preferably a positive consequence). Edward Feeney introduced the concepts of behavior modification to industry in the early 1950s. In fact, he produced training materials with B.F. Skinner, showing how behavior modification concepts could be used to increase productivity, reduce customer complaints, and reduce absenteeism in an organizational setting.

All of this happened long before safety had even heard about such things—even long before the early Bird and Schlesinger article.

Safety practitioners have over the years done some work and thinking about behavior as a part of the safety problem. H.W. Heinrich suggested in 1931 that 88% of all accidents are caused by unsafe acts of people, pointing out that we must focus on behavior in order to control accidents. We all agreed, but chose to do little about behavior except to train and discipline, ignoring the available knowledge from Watson, Pavlov, and Skinner.

A few behaviorally sound approaches were used as early as the 1960s. In 1962, in his article "Safety Sampling," Vincent Pollina showed us how to measure behaviors using quality control concepts. The technique had been successfully used at that time in heavy manufacturing plants. (It is seldom used today.) In 1968, I published an article in *Professional Safety* discussing a management model developed by Edward Lawler and Lyman Porter in 1967. That article describes clearly that the safety performance of managers and supervisors is determined by how their roles have been defined, how they are measured, and how they are rewarded.

During the 1960s, we had full access to the psychologists' research in humanistic approaches for getting desired behaviors. Argyris showed that meeting the mature, adult human needs of people in fact influences behavior positively. Herzberg suggested that most people respond to achievement, recognition, responsibility, and the other identified motivators. Few in safety used this knowledge, but it was a commonly known fact in industry. We also had full access to the behaviorists' knowledge that behavior is modified by the consequences of that behavior.

In 1971, my *Techniques of Safety Management* was published, covering some of this. In 1971, OSHA regulations also went into effect, pushing safety management concepts back down the unsafe condition path and putting off the application of behavioral approaches for many years. In 1975, I published *Safety Management—A Human Approach,* on both the humanistic and behavioristic approaches to obtaining safe behaviors.

The point of all this history is that the fundamentals of behavior-based safety have been with us for almost 90 years; we have used them for 50 years in industry and for over 30 years in safety.

Today behavior-based safety is popular; it is discussed at every safety seminar, symposium, and conference. It is often presented as new knowledge, as new concepts. Psychologists have become workplace consultants, with some using culture-building approaches, while others are concentrating on employee observations.

Neither approach is new. Although some advertise and market the concepts heavily or even copyright the terms, they are still old concepts. Organizations today will pay a fortune for behavioral packages, but they can get the same information and the same approaches from a freshman-level college textbook.

Much of the current behavior-based safety management is excellent and long overdue in safety, but some has become counterproductive—even dangerous.

It is excellent and long overdue that we are finally looking at and using the knowledge of the past. In some cases, we are measuring and building cultures, finally understanding that the culture of the organization as perceived by the employees literally dictates which elements of the safety system will work and which will be a waste of time. In other cases, we are finally concentrating on Heinrich's 88% (unsafe behaviors), observing them and intervening. We should have been doing these things from the beginning. When Heinrich said (whether factual or not) that 88% of all accidents are caused by unsafe acts, we should have asked Watson, Pavlov, and later Skinner, what can we do to lessen the probability of these acts occurring? They would have told us. Regardless of the lost years, to put the focus on behavior now is truly a breakthrough for safety, and should be a wave for the future.

What's not so great and perhaps even counterproductive is that, in many organizations, approaches that emphasize peer observations have actually shifted the entire safety effort to the hourly employees, allowing easy management abdication. In all the research and benchmarking that has been done, it is clear that primary emphasis must remain on management involvement. Management is the key.

In some cases, behavior-based safety is perceived as "the safety program," allowing other crucial elements to wither. In other cases, the results of peer observations are never measured. They simply don't know if any behaviors have actually improved. Or they measure the number of observations generated without a valid measure of the reduction of unwanted behaviors.

In the companies that I have worked with in recent years, I see more paperwork generated with fewer results than before the behavior-based era. I also see a wider chasm between management and the work force, with management further out of touch than ever before. These changes are dangerous. Safety can only be achieved when both management and employees do it together, when there is confidence and trust existing between them, and when both perceive safety to be a value, not a priority that is subject to shifts depending on what day it is.

Another trend that is counterproductive is that we seem to have shed the old safety mythology (which management never really understood) for a new behavior-based safety mythology (which management doesn't understand). We use different terminology for much of the same old stuff, as if it is new, wonderful, and still mystical. In some cases, the safety rules, safe operating procedures, job safety analyses, and job safety observations that evolved over many years of experience are thrown out, and critical behavior inventories are developed. Often such inventories contain only the most frequently occurring behaviors, ignoring those less frequent behaviors that can lead to severe injury. We've known for years that the causes of the severe, the catastrophic, accidents are different from the causes of the minor and frequent.

Sometimes, we spend months training people to do things they've always known how to do—taking them off the production floor, forcing the remaining people into even more work, more forced overtime. In some organizations, instead of the traditional two tiers (management and workers), we see three (management, involved workers, and noninvolved workers), with as much friction between each as used to exist between the former two tiers.

WHERE ARE WE HEADING?

Many organizations are beginning to realize that the downsizing and pressure-building, stress-causing management concepts have gone too far. Some organizations are beginning to add people; some are thinking of culture building. At the same time, cost containment remains critical, and many initiatives of the organization have to show that they add value.

Robert Kaplan, of the Harvard Business School, writes of the necessity for any organization to look critically at any product or function and to consider the following tests:

1. Identify externally focused measures of how the product (or program) gives value to the customer and how it supports the goals of the business. In safety, this clearly requires totally new safety and accounting measures.
2. Identify activities that cause work on the product line and assess how each activity adds value or creates waste. A listing of activities enables us to search for the type of activity that consumes resources but adds no value. In safety, we probably will have many.
3. Identify and eliminate the generators of work that cause nonvalued activities. Could any of these be safety activities?

It would seem that any element of a safety system should be subjected to the above three tests, including the behavior-based functions. Good companies will do this; and eventually most companies will be forced to. This will affect the future of behavior-based safety. If the behavior-based system does, in fact, add value, it will remain. If it creates waste, it will not.

To determine this, we will need externally focused measures (statistically valid safety sampling upstream, perception surveys and reduced costs downstream). Some programs will survive and some will not. If behavior-based safety is perceived as a program, its future is dead. If it is perceived as the way we manage, it will survive. If it reduces the chasm between management and the work force, it will survive.

In the organizations I'm familiar with that have installed behavior-based systems, I'd predict that perhaps less than half would survive these tests. Some stand out like shining stars where behavioral approaches (which usually include both humanistic and behavioristic approaches and which always include both management and the hourly work force) are an integral part of the way the organization is managed. Other behavior-based systems remain in the memory of all, since they no longer exist. And others may no longer exist if subjected to Robert Kaplan's three tests.

REFERENCES

Bird F, Schlesinger L. Safe behavior reinforcement. *Professional Safety* June, 1970.

Heinrich H, Petersen D, Roos N. *Industrial Accident Prevention*, 5th ed. New York: McGraw Hill, 1980.

Kaplan R. *Measures for Manufacturing Excellence.* Boston: Harvard Business School Press, 1995.

Lawler E, Porter L. *Attitudes Affecting Line Managers' Performance.* University of California: Institute of Industrial Relations, 1967.

Petersen D. Attitudes affecting line managers' safety performance. *Journal of the ASSE* May 1968.

Petersen D. *Techniques of Safety Management.* New York: McGraw Hill, 1971.

Petersen D. *Safety Management—A Human Approach.* New York: Aloray Publishers, 1975.

Pollina V. Safety sampling. *Journal of the ASSE* August, 1962.

Skinner BF. *Science and Human Behavior.* New York: McMillan, 1953.

Watson J. Psychology as the behaviorist views it. *Psychological Review* 1913.

2 Early Attempts at Employee Involvement

Attempts to involve employees in the safety management process have a long history. The participative approach was suggested as early as 1937 by H. Carey, who defined consulting supervision as the procedure whereby supervisors and executives consult with employees or their peers on matters affecting employees' welfare or interest prior to establishing policies or initiating action. From such beginnings, the concept of participative management has grown into a full-fledged approach to administration, affecting both patterns of organizational leadership and leadership style. Participation in Carey's sense can be construed as a comfortable rationale for the paternalistic manager. Applied in the modern interpretation, it complements the political and social philosophies of democracy and individual self-actualization. Today we believe that participation enhances the learning process and that a democratic environment is more conducive to productive effort.

Participation Experiments

A few of the best studies of participation show how the application of this technique can be used to further the adjustment of workers to changing conditions. The best-known early studies are those that were carried out in the Harwood Manufacturing Company. In this company an attempt was made to come to grips with the workers' reactions to retraining. Harwood had especially severe turnover problems among workers forced to transfer from one job to another because of changing manufacturing needs. Many of these workers actually quit shortly after the retraining had begun. There were also a large number of voluntary quittings during the period just before the workers reached factory standards of competence at the new jobs. In addition, retraining was slow and costly. Transferees took much longer to learn new jobs than did new workers, even those who had been highly skilled at their previous jobs.

A controlled experiment was set up in which one group of workers about to transfer was given routine information about the nature of the change and nothing else. Three other groups were asked to plan the way in which the factory would adapt to shifts in the market and in technology. These latter groups arrived at the necessity for their own retraining by means of group discussion. The group that had merely been ordered to retrain showed high turnover and a long training period. The groups that had discussed the shift and arrived at their own decision about the necessity of retraining showed virtually no turnover and a phenomenally shortened training period. At a subsequent period, the original "no participation" group was reassembled and again retrained. This time, they participated fully in the decision-making process. Figure 2-1 shows the retraining response differences of this group when they did participate and when they did not participate in planning the shift.

Participation has been found to be successful in many situations. It does not always increase productivity (sometimes it increases, sometimes it stays the same), but it does always improve attitudes, turnover, morale, etc. Workers do want to be involved in decision-making where they feel they have a right to be involved. Workers' attitudes toward both company and union are strongly affected by the degree to which foremen and shop stewards welcome the participation of the rank-and-file workers in arriving at decisions. Involvement of all levels of employees in decision-making is a major factor in morale. In fact, the idea that it is a good thing to get employee participation in making changes has become almost axiomatic in management.

Participation, however, is not something that can be created artificially. We cannot buy it. We cannot hire industrial engineers, accountants, and other staff people who have the innate ability to "get participation." We cannot tell our supervisors and staff persons to just start participation.

People respond to the way they are usually treated, rather than to a sudden strategy of participation. Many supervisors have had unhappy experiences with executives who read about participation and pick it up as a new psychological gimmick for getting other people to do what they want. Participation

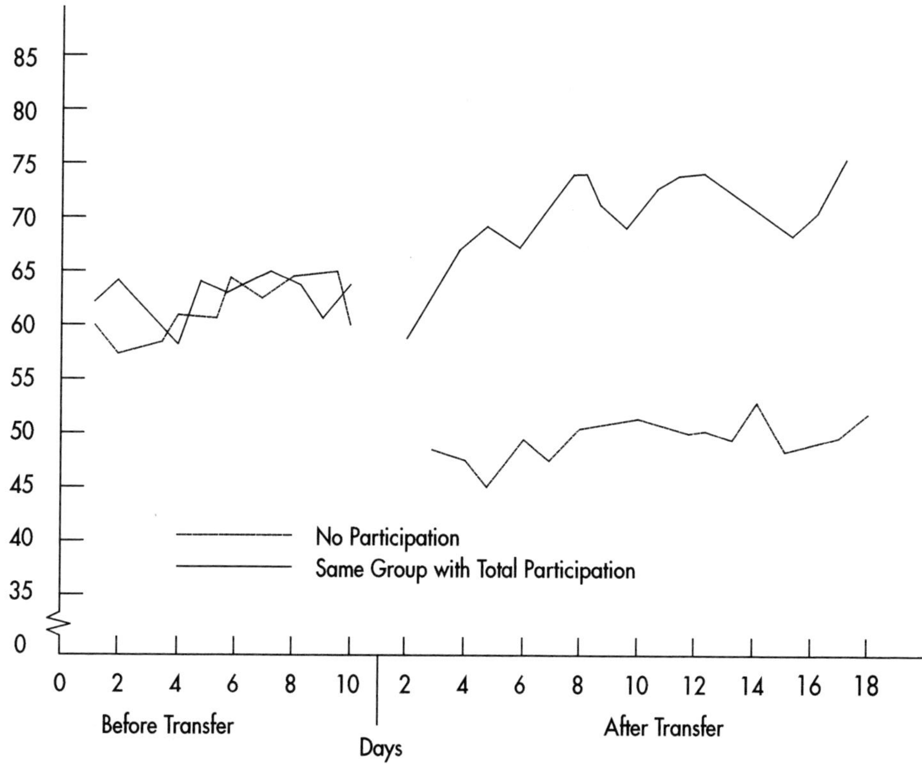

Figure 2-1. A comparison of the effect of "No Participation" with "Total Participation" on the same group.

Source: D. Petersen, Safety Management—A Human Approach, *2nd ed., New York: Aloray, Inc., 1988, p. 195.*

results from a feeling that people have, not just from being called in to take part in discussions. Participation must be based on respect, and respect is acquired when management faces the reality that it needs the contributions of the operating people.

Safety professionals believe that participation improves the safety record. One of the main proponents of participative safety was Robert Burns (General Electric), who said,

> *One of the problems inherent in the traditional way that safety rules are formulated and administered is that the few at the top of the organization dictate what is proper or improper safety-wise for those at the bottom. The job of the supervisor is to enforce the rules, and his subordinates are to follow*

them. Since most people resist being told what they can or cannot do, each rule becomes a restriction of one's personal liberty. Acceptance of restriction is difficult for most people, unless they are convinced of the validity of the restriction. For example, most people stop for red traffic lights because they are aware of the validity of a rule which prevents accidents. Without acceptance, any rule or law will either be ineffective or it will cause problems in enforcement.

If we admit that there must be a system of rules in order that an organization function effectively, then it is necessary for industry to find an imaginative approach to a system which will initiate and develop a solid set of safe-practices specifications which will be accepted and enforced as valid guides to safe behavior by those who must work within their bounds. Acceptance is the key requisite and acceptance hinges on the obvious validity of any rule. One of the basic principles of nondirective counseling is that an individual tends to accept his own conclusions. A group will tend to accept its own conclusions as well. Positive acceptance of safety rules can result if subordinates suggest and develop their own safe methods and rules. After all, who knows more about the work and its hazards than the people actually doing the work? Motivation to follow the safe practice, then, would come from within each person as a part of the group, and is not the result of outside pressure. Once the people of the department decide what rules of safety should apply to the activity of their department they will, when necessary, enforce their own rules much more effectively than can the supervisor under the traditional punitive system.

In a positive approach towards safety, the role of the supervisor becomes one of key importance; he becomes the leader in helping the group develop its rules of conduct. The role of the safety director becomes that of the expert to whom the group looks for professional help and advice. The safety director and the supervisor listen and encourage the members of the group to express their ideas. They ask questions in order to be sure that all sides of an issue or method have been considered by the group, and they supply necessary technical information when it is required. They function as they should—as leaders, not tellers. With the help of the leaders, the departmental group is able to develop standards of behavior that will readily meet the needs of the department. Those who might refuse to do things the group's way, because they think their own way is best, can pretty well be persuaded to conform in the areas of safety under the threat of ostracism. Since the need to belong to the group is a powerful motivator, the potential rule-breaker will probably tend to conform.

Another benefit from this approach to rule-making is that it gives a number of people, perhaps for the first time in their lives, a degree of participation in directing the activities of the department. This can create a sense of involvement and commitment to the goals and the objectives of the department—goals and objectives that they have a part in establishing. From the area of safety, this involvement and committment can spread into many other facets of the department, permitting each member of the organization not only to contribute to his maximum, but also to achieve greater job satisfaction than ever before.

Source: R. Burns, "Participative Safety: A Motivating Factor," *ASTME Vectors*, vol. 5, 1969, as quoted in D. Petersen, *Safety Management—A Human Approach*, New York: Aloray, 1975, pp. 93-96.

Participative safety programs have been tried and they work. One Midwestern company had the following results:

As this program developed employees recognized that they had a stake in safety, and we—the management—stayed away from telling them how to do it. Then, instead of giving lip service to our efforts to enforce rules, employees reacted positively because the program for safety was theirs.

Another stated:

Physically, little has changed; the machines, the protective devices, the products, the procedures, and most of the employees are the same as they were 18 months ago. What has changed is the attitude of the employees. Today, they are concerned with the safe method of performing the job, because they gain individually and collectively by being safety conscious. They also are aware that only their efforts can prevent accidents.

And the workers agree:

We had a safety committee for six years, but nobody took it to heart. Now we're thinking safety and quietly anticipating hazards before they create accidents. Some of the older employees make a definite effort to get any new fellows in on the program and watch their work on the machines to head off anything before it happens.

PARTICIPATION AND DISCIPLINE

One of the real advantages of participative safety is that it almost eliminates the need for discipline. Robert Burns states,

One of the main problems inherent in safety is that there is no particular satisfaction gained from working in a safe manner: you may avoid an accident by working in a safe manner, but you never really appreciate an accident that is avoided, because you are usually unaware of it. While the motivation to work safely is vague, the motivation to violate safe working practices is more concrete: the prescribed method is inconvenient; it requires more effort; it takes time when one is in a hurry; there may be a fear that coworkers may think one is timid, or afraid if he follows all the safety rules; perhaps dislike for the boss or the company is enough reason to ignore safety regulations. When there are more motivating factors working against safety than for safety, the tendency of workers will be to follow the unsafe method.

Once workers have been trained in safe working methods, the problem of motivation—usually the responsibility of the supervisor—becomes primary. Most supervisors approach the problem of motivating safe work practices by means of the use of punishment. In punitive systems, the subordinate is presumably motivated to follow the safety rules because he is afraid that punishment will follow if he does not. A rule with a set penalty for infraction is an example of the fear punishment method. While it is true that in many cases this method is effective, in others it has a negative effect.

Punishment has negative implications: it may result in wrong learning, that is, learning how to avoid getting caught, and it can create hostility within the worker when he is caught. Fear can motivate people to avoid unsafe methods, but this avoidance does not necessarily lead to an acceptance of the safe method. It is a negative emphasis rather than positive, and, as a rule, the gains realized by negative motivation are short-term. As a negative force, punishment can create hostility: a schoolchild who is punished for fighting may fight even more in the future, because of the hostility which can develop as a result of the punishment, and hence, further infractions, may result. In the factory or the office the behavior of a subordinate who is punished for breaking a safety rule can become hostile in the same manner as the behavior of the youngster punished in school.

Punishment makes impossible the development of a strong, positive relationship between the subordinate and his supervisor because it is negatively oriented. In fact, it can encourage a deterioration of the company-supervisor-subordinate relationship. In companies where policy calls for a disciplinary layoff for infraction of safety rules, supervisors may take great pains not to see a violation in order to keep themselves from being placed in the position of administering punishment. This "voluntary blindness," in effect, implicitly encourages the unsafe method. Since safety rules cannot be eliminated—since they are mandatory in every facet of industry—the problem is one of how to increase the positive influences of safety rules, while minimizing negative reactions by subordinates.

Burns' views on participative safety and on punishment tie in closely with what the psychologists now know about punishment. Punishment following an act tends to teach a person not to get caught in the act again, rather than not to perform the act at all. In safety, this is not good enough. We must teach people not to perform the unsafe acts whether or not management, or the supervisor, are observing them. Participating in the decision as to what is an unsafe act is the point where participation leads to the behavior we want. Having participated in that crucial decision, the employee no longer is a discipline problem to us. As Burns summarizes,

If industry cannot eliminate safety rules, then we, as leaders in industry, should abolish the traditional methods of establishing them. Part of the rule-setting responsibility belongs to those people for whom the rules are intended. Under the direction of a safety director and his subordinate, the supervisor-leader, positive motivation towards safety rules and positive enforcement of safety rules can result, first, in sound interorganization, and second, in sound economics: fewer lost man-hours, less compensation for injuries sustained on-the job and less replacement of skilled help. Each means gained dollars.

Source: R. Burns, "Participative Safety: A Motivating Factor," *ASTME Vectors,* vol. 5, 1969, as quoted in D. Petersen, *Safety Management—A Human Approach,* New York: Aloray, 1975, pp. 93-96.

IS PARTICIPATION FOR EVERYONE?

Roderick Forsgren (University of Maine) made a detailed study of participative safety programs and identified the conditions needed for participation to work. They include the following:

There are some very basic conditions which must exist before individual employee participation in safety programs becomes an effective technique. A person must be free to rearrange his particular goals in light of his experiences. Involvement in safety programs must provide experiences which allow a person to alter his beliefs and sentiments concerning safety as it relates to his work.

Individuals must also be willing to accept participative activity before it becomes an effective managerial technique. People with low independence needs, people who work under safety programs that are highly authoritative and reflect central control might well find opportunities for participation as signs of weakness in leadership. Thus, hopes for immediate acceptance and compliance via employee participation may be dysfunctional in light of the safety objectives.

Workers in most organizations are, however, accustomed to some degree of participation, although it may be at a very low level. There is probably far less danger in extending participative techniques in the safety area than is imagined by many safety directors.

Participation in safety programs must also be viewed by the individual employee to be relevant and important. His involvement in safety programs should be perceived by him to increase positive goals and reduce negative ones. It might well be that these positive goals extend beyond the realm of safety and include such factors as recognition, status, power, and achievement.

Thus, the use of employee participation in safety programs carries with it the "risk" of permitting employees to make decisions in areas that management may have traditionally considered its perogative. Changes in work-flow or work-methods design that may have safety implications would appear to be an example of a "hot" area; whereas extending participation to areas which are judged irrelevant by the employee result in very little or no return to either the individual or the company.

A final criterion for effective participation in safety programs is the belief by employees that the area in which their decision-making is permitted is one which they view as legitimate. By legitimate is meant an area in which the employees feel it is right and proper to engage in the decision process.

Although there may be some exceptions, participation in safety programs generally does not offer situations in which individuals doubt they have a right to make decisions, for very often, unsafe conditions are directly related to work and almost always viewed as affecting a person's safe being.

One must be careful, however, not to confuse legitimacy in safety with importance of safety. These two concepts are not identical. It may well be because almost all safety practices can be interpreted by the individual to have some direct bearing on his well being, and, hence, he may view it as a legitimate area. Whether or not he considers it important is a separate matter.

Thus, in summary, the approach by line personnel in implementing safety programs is dependent upon four basic assumptions which must exist before participative techniques are useful. These are (1) psychological involvement, (2) fulfillment of a need for independence, (3) the relevancy or importance of the area of decision making, (4) the extent to which the employees view the area as a legitimate one in which to make decisions.

The absence of any one of these four factors makes the implementation of safety programs through the line organization via employee participation an ineffective venture.

In addition, certain organizational requirements are necessary. For example, the use of participative techniques must be economically feasible. Admittedly, it is virtually impossible to determine the exact costs of participation. However, there are times when the costs of the participant's time, for example, would be prohibitive. There are practical restrictions in addition to time and sheer cost, including employee experience and adequate communication structures, which can severely restrict the organization's use of participative techniques.

Source: R. Forsgren, "A Model of Supportive Work Conditions Through Safety Management," *J of ASSE*, Feb. 1970, as quoted in D. Petersen, *Safety Management—A Human Approach*, New York: Aloray, 1975, pp. 200-203.

Participative safety casts each worker into a role quite different from the role in a hierarchical safety program. The participative program changes the worker from an inferior subordinate to an equal, a peer; from a person who receives an order and does the work as ordered to one who helps decide what is to be done and then acts in accordance with those decisions. This molds new values, attitudes, and changes in behavior. In short, under participative safety, the worker becomes a motivated individual because the role is that of a person who wants to accomplish some specific results. The decided goals are group goals, not goals thrust upon them. And since the group has set the goals, achieving those goals is accepted; it is the group norm. Behavior that leads to group goal achievement is normal behavior. Behavior that does not lead to this becomes deviant behavior. A perfect example is in the wearing of personal protective equipment. Thirty years ago, hard hats were virtually unknown in the construction industry. Today, practically every construction worker wears one at all times. The hard hat has actually become a symbol of the industry. Before hard hats were accepted, they were considered a taboo item to wear. Anyone wearing a hard hat was considered a sissy on most jobs; the wearer was the deviate. Today the individual construction worker is almost forced into self-protection from serious injury. To not wear a hard hat is deviant behavior today.

John P. Gausch advocates striking a balance between authoritarian safety programs and participative programs. His study of fifteen locations with 10,000 employees resulted in the following conclusions, which he labeled as opinions:

- *"It is not supported that an expression of strong participation—self-control, group decision making, and reduced authority—will assure any improvement or be related positively to safety experience.*
- *It is not supported that an expression of strong authority—warnings, threats, reprimands, and reduced participation—will assure any improvement or be related positively to safety experiences.*
- *It is supported that an expression of strongly balanced involvement tends to have a respectable validity coefficient with improved safety experience.*

He further offers the prediction model shown in Figure 2-2.

Dr. M. T. McDowell, President of McTar Petroleum of Vancouver, B.C., Canada, described some case studies of safety programs that utilized participation:

> *The two cases summarized are from very different industries (construction and manufacturing). The studies sampled are longitudinal in nature (3-5 years). In both cases the Traditional Safety Program was replaced with a program involving some contemporary behavioral science wisdom (the Modern Safety Approach). For both industries the introductory process as well as content and format of the Modern Safety Program was identical. It should be noted that Safety and Accident Prevention is used by the author as an entry vehicle to conduct larger Organizational Development change interventions.*

How Traditional Safety Programs Differ from Modern Safety Programs

Traditional Safety Programs Typically Have:

1. *One Head Office person responsible for Safety*
2. *An organizational safety specialist usually referred to as a Safety Director (Sometimes Safety Directors have Departmental support staff.)*
3. *Safety as a staff function with director and department as staff positions*
4. *A set of specific Do and Don't Company Safety regulations set down by management*

CONTENTS OF PROGRAM

Strong Participation	+	Strong Control	=	Result
No	+	No	=	Poor
Yes	+	No	=	Questionable
No	+	Yes	=	Questionable
Yes	+	Yes	=	Good

Figure 2-2. Prediction model

(Reprinted with permission from J. P. Gaush, "Balanced Involvement," *Monograph 3,* DesPlaines, IL: ASSE, 1973.)

5. *Marginal line commitment to safety and moderate involvement in accident prevention.*

The Modern Safety Approach Has:

1. *On the job worker and work group responsibility for accident prevention and safety goals*
2. *Organizational safety specialist usually referred to as a Safety Coordinator*
3. *Safety and accident prevention as a line responsibility of line organization*
4. *Emphasis on upward communication from work group to supervisors to middle management to top management. Also lateral communication to the Worker's Union and the Worker's Compensation Board as well as between work groups.*
5. *Umbrella policies from management protecting worker input as well as specifying structure of committees, nature of communication flow and reward system. These policies focus on the process of the Safety Program. The exact content of the rules and regulations is left to the particular work groups to formulate and enforce (The Buddy System).*
6. *Management and work group involvement in and commitment to accident prevention goals and safety programs.*

Source: MT McDowell, "Current Content Issues on Organizational Change—Safety," unpub. paper, 1982, as quoted in D. Petersen, *Safety Management—A Human Approach,* 2nd ed., New York: Aloray, Inc., 1988, pp. 157-160.

Safety has a rather unique quality relative to all the other organizational dynamics. It is pervasive in its importance throughout the organization. This aspect of safety affords the change agent instant organization-wide visibility while at the same time allowing him or her to enter the system on an important nonthreatening issue. A successful impact on safety performance and worker well-being can have a lasting and profound impact on other improvement efforts. The consultant is as legitimate in the executive office discussing safety with the president as he or she is on the floor or in the trenches with the work group.

The central thrust of the modern approach to safety and accident prevention is the committed involvement of the intact work group. Workers in traditional settings often report a sense of powerlessness in their jobs. That is a feeling of being unable to influence policies, procedures, and conditions. The resultant attitude is often verbalized by a conclusive comment like "that's the way it is and will always be." This attitude often comes from years of confirmed experience. By placing safety and accident prevention directly in the hands of the workers and work groups the "ever thus" attitude can become a "yes we can" commitment. The involvement in safety can overflow into decision making and problem solving around other aspects of the work environment.

QUALITY IMPROVEMENT TEAMS

One of the popular participation techniques is the use of quality improvement teams (QITs), which started in Japan in the early 1960s and about ten years later found their way to the United States. A QIT consists of a small group of workers who do the same type of work and have formed together under management's auspices to solve daily work-related problems. They are purely voluntary groups that meet on a regular basis and select their own group members and problems to solve. Often, they use sophisticated statistical tools to help solve these problems. They then submit their analysis and suggested solutions to management for decision and implementation.

Figure 2-3 shows how the QIT concept fits with other theories discussed in this book. Usually an organization will first form some type of structure to facilitate participation in the QIT. As shown in Figure 2-4, this structure often mirrors the organizational structure. Often the QIT concept is installed slowly, perhaps in a pilot plant or a pilot section of the plant, and allowed to grow on its own. Typically, the growth is fast because QITs are popular in most organizations.

THEORY OR PRINCIPLE	COMPATIBLE QIT CONCEPTS
"Scientific Management" *(Taylor)*	Quality circles place an emphasis on statistical tools and work analysis.
"Hawthrone Effect" *(Mayo)*	Special attention is available to circles via presentations to management and communication of results.
"Hierarchy of Needs" *(Maslow)*	The structure and processes of QITs help employees to meet higher-order needs, particularly those dealing with social belonging, esteem, and self-actualization.
"Theory of Personality and Organization" *(Argyris)*	Mature, responsible behavior is the essence of circle membership.
"System 4" *(Likert)*	Quality circles are a tool for achieving a participative group approach to management.
Power, achievement, and affiliation motives *(McClelland)*	Quality circles provide opportunities for satisfying all three motives through autonomy, accomplishment, and social interaction.
"Motivation/Hygiene Concept" *(Herzberg)*	Quality circles supply the motivational factors of growth and learning, recognition, achievement, the chance for advancement, and an improved perception of the work itself.
"Theory X" and "Theory Y" *(McGregor)*	The managerial attitude on which QITs are based consists of respect, trust, and belief in the capability of workers to solve problems, create ideas, and take responsibility for their own behavior.

Figure 2-3. Management theories and compatible QIT concepts.

Training in the problem-solving process, and often training in the use of statistical methods in problem solving, precedes group installation. Although you don't need QITs to get participation, they have been found to be highly successful.

Obviously, not everyone agrees wholeheartedly with participative safety. Safety professionals are showing cautious interest conceptually, but few are doing much yet with the concepts. No proof exists that participation will definitely yield a lower accident rate than nonparticipation, so few try it. Nor does proof exist that authoritarian safety programs will yield a lower accident

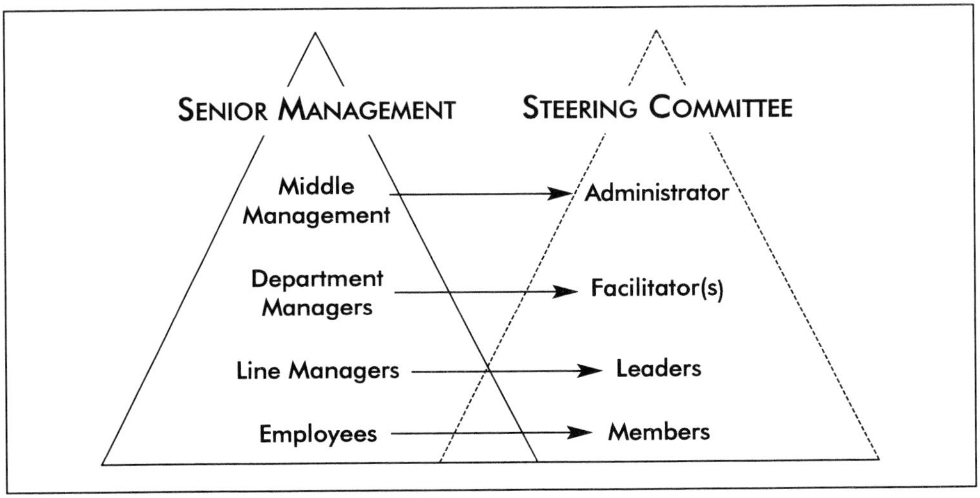

Figure 2-4. Comparison of organizational hierarchy and QIT hierarchy.

rate than participative programs, but it is here and we are comfortable with it, so mostly we stay with it.

H. Harvey Cohen, writing in *Professional Safety* magazine, offers advice on implementation of a participative safety program:

> *Considerable preparation is necessary for all taking part in implementing such programs. This includes not only technical personnel handling the implementation, such as facilitators and higher management, but all individuals who participate.*
>
> *Lower and middle management need to be prepared in several ways. First, they must be trained in their new and different roles, especially vis-à-vis subordinates who have increased roles in decision making. This will make their adjustment easier, leaving them with less of a feeling of lost power and importance, and will help them to adapt to their new roles and responsibilities.*
>
> *Second, managers at all levels must really believe that their subordinates have something to contribute, otherwise, employee attempts to provide input will not meet with much acceptance or interest.*
>
> Source: H.H. Cohen, "Employee Involvement," *Professional Safety*, June 1983, as quoted in D. Petersen, *Safety Management—A Human Approach*, 2nd ed., New York: Aloray, Inc., 1988, pp. 161-163. Reprinted with permission.

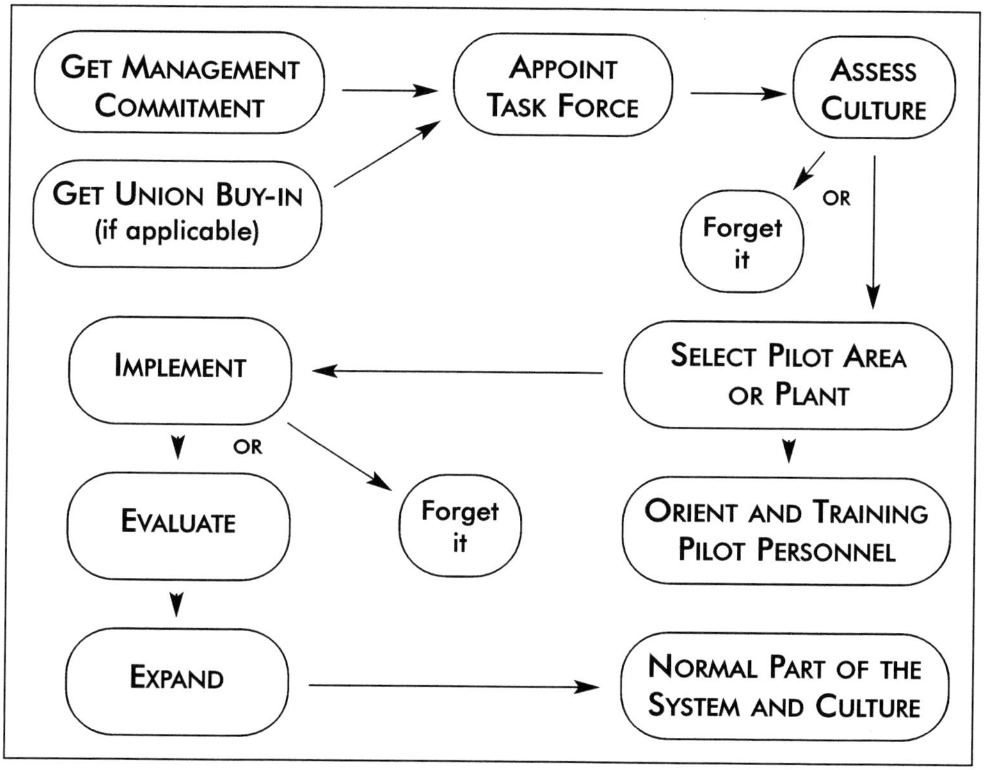

Figure 2-5. Quality improvement team installation process.

Employees, too, must be prepared. They need training in new techniques, new work patterns, and the like. Less obviously, they need guidelines in areas of participation. For example, if problem solving is an important new part of employees' jobs, then training in problem-solving methods is a must. If this will be done by committee or group, training in effective interpersonal communication is essential.

Implementation of an employee involvement program (or, for that matter, virtually any organizational restructuring) takes time. Immediate positive results should not be expected, and in fact, some grumbling and resistance is inevitable. (One goal of the preparation discussed above is to reduce such resistance by increasing understanding of the new methods.) It takes time for workers and management to become accustomed to new ways of doing things and to take maximum advantage of them (see Figure 2-5).

ARE COMMITTEES PARTICIPATION?

In the safety programs of the past, the closest thing we had to real participation was the safety committee. The pros and cons of safety committees have been aired for years without changing anyone's mind about them. They exist, and in some companies they fulfill a role of communication, quasi-participation, or perhaps in some cases, real participation, for the few workers on the committee. The safety committee, however, is not what we are talking about when we talk about participative safety. We are talking about employees participating in the safety decisions that pertain to their jobs.

REFERENCES

Burns R. Participative safety: A motivating factor. *ASTME Vectors.* Vol. 5, 1969.

Carey H. Consulting supervision. *Nation's Business.* April, 1937.

Cohen H. Employee involvement. *Professional Safety.* June, 1983.

Forsgren R. A model of supportive work conditions through safety management. *Journal of the ASSE.* February, 1970.

Gaush J. Balanced involvement. *Monograph 3.* DesPlaines, IL: American Society of Safety Engineers, 1973.

McDowell M. Current content issues on organizational change—Safety. Unpublished paper, 1982.

3 Rationale for Involvement

During the last ten or fifteen years, management has undergone some amazing and rapid transformations. As each new book is published, management changes its direction and philosophy on how to manage.

The transition in management thinking stated a number of years ago when managers began to question the dictates of *classical management or scientific management* as described by Frederick W. Taylor.

Following the classical school of management thinking came the *human relations* school of thought. To a degree, the human relations approach, popular in management until the 1970s and still taught today, was an outgrowth of the misinterpretation of the research of the 1920s, '30s, '40s, and '50s. Many interpreted the research to mean that happy workers are productive workers, thus the role of management was to make workers happy. Actually the research did not say exactly that, as it is considerably more complicated than that.

The underlying assumptions of the above two schools of management thought were similar but slightly different. Classical management is based on the assumption that "People are all alike—we can get the behaviors we want through manipulation." Thus we used wage incentive plans, piece rates, and other schemes to get more productivity. Human relations management has the underlying assumption that "Everybody's alike—we can get the behaviors we want by making them happy." Neither assumption has any basis in fact.

In the 1970s the *contingency school of management training* emerged, and for the first time, management's underlying assumptions about people were in tune with psychological reality. The assumption behind the contingency theory of management is that "everybody's different"—so our management style, and how we deal with workers, must be contingent upon the situation, upon the workers, and upon their needs. How a manager manages must be appropriate to the situation.

Until the 1990s, the philosophy of safety management remained fairly comfortable, for we could still try to feed our classical management style safety programs into our organizations. But in the 1990s, new and different trends in management became popular. The Deming philosophy and total quality management concepts began to threaten our comfort levels, for now we were talking employee empowerment, with no rule-setting by management, using statistical process control (SPC) tools that could literally make many of our old safety tools obsolete. In addition, rightsizing and reengineering in some companies had the effect of putting more and more work on fewer and fewer supervisors, managers, and employees.

In this evolutionary process, one of the emerging trends has been to look at high-performance organizations, and decide that some levels of the traditional hierarchy were unnecessary. In some organizations, this has meant a flattening of the organization, reducing the number of middle managers and the number of layers in the hierarchy. By flattening the structure, the organization should become more efficient, more productive.

In some organizations, first line supervisors were found to be unnecessary, and team management was established. This decision has great historical underpinnings—the Hawthorne studies of the 1920s showed that the most productive work units are those with no formal supervision. It should be remembered, however, that in those studies, team leadership was only an experiment, and, as a result, people responded very positively (the Hawthorne effect).

As we go from supervisor control to team control, how do we ensure an accident/illness-free operation? The team approach requires us to reassess and reunderstand how safety can be accomplished in any organization.

In safety, with rare exception, we do not use psychological theories, i.e., what is known in motivation theory or in behaviorism concepts. We replace motivation theory about understanding and satisfying the worker's needs with training, contests, gimmicks, and awareness campaigns.

In behavioral concepts, we do equally poorly. We turn over the whole process to the work force, allowing supervisors and managers to abdicate. Instead of positively reinforcing after safe behaviors and attaching the reinforcement to a specific behavior, we reinforce sporadically, or at random, and attach it to no particular behavior.

BEHAVIOR-BASED SAFETY APPROACHES

In recent years, the safety world has been informed that our future progress will come from behavior-based approaches. The difficulty to many of us in the field is that although we tend to agree, we're not quite sure what behavior-based means. We've believed since the 1930s that most accidents are caused by human behavior (unsafe acts) and that the preponderance of our efforts should focus on lessening the probability of unsafe behavior. Does this mean that most of what we have been doing traditionally can be called behavior-based approaches? In the 1970s, regulations and fear of fines forced us to concentrate on physical conditions, health, record keeping, machine guarding, occupational diseases, ergonomics, etc., depending upon the year and the current emphasis from the government. Probably defining those thrusts as behavior-based would be a stretch, but training, awareness campaigns, use of media, continued during those years in most organizations that did aim at influencing the worker's behavior.

For the last ten years or so, behavior-based safety has been very popular. Is it the elusive magic pill we have been searching for throughout our history, or is it the same stuff we've almost always done under a different title?

Perhaps the most frequently heard definitions of behavior-based safety describe it as a process of involving workers in (1) identifying the ways they are most likely to get hurt, thus getting their involvement and buy-in, and (2) asking the workers to observe other workers to determine progress in the reduction of unsafe behaviors. Is this all there is to behavior-based safety?

This picture, if accurate, suggests the need to go back to the basics of psychology and the fundamentals of behavior. I learned that we have three basic ways to influence behavior:

1. to change attitudes in the belief that our behavior is consistent with our attitudes
2. to build a psychological climate in which people will choose to behave as we wish, since that behavior helps to satisfy their current needs
3. to modify behavior through the systematic use of reinforcement following a behavior to influence future behavior (called behavior modification)

Traditional safety was #1 and #2 above. We try to change attitudes and use motivation concepts ahead of time to influence future behavior.

Over the years we have tended not to use #3, except the occasional use of discipline (negative reinforcement) following an unsafe act or an injury. True behavior-based approaches would use #2 and/or #3, motivation theory and behavior modification, and not #1, attitude change. Attitudes are impossible to measure, and changing attitudes is complex. Measuring behaviors is easy and influencing or modifying behaviors is possible through a number of different options. The following sections discuss the two options we have to influence behavior.

BUILDING THE PSYCHOLOGICAL CLIMATE

Using motivation concepts to change behavior has been around for a long time in organizations and in safety. Motivation has to do with satisfying the current needs of the individual worker. It does not have to do with training, awareness campaigns, the use of media, or gimmicks. Motivation takes place when the organization, or the boss, understands the worker's needs, and creates an environment where those needs can be met on the job. What are the needs? Old and current research and theory spell them out.

Conflict Theory

Bringing together the evidence regarding the impact of formal organizational principles on the individual, Chris Argyris, a major psychologist, concludes that there are some basic incongruences or conflicts between the growth trends of a healthy personality in our culture and the requirements of a formal organization. The individual in our culture tends to develop from childhood to adulthood along specific trends as follows:

- from a state of passivity as an infant to a state of increasing activity as an adult
- from a state of dependence on others as an infant to a state of relative independence as an adult; relative independence is the ability to "stand on one's own two feet"
- from being capable of behaving in only a few ways as an infant to being capable of behaving in many different ways as an adult
- from having erratic, casual, shallow, quickly dropped interests as an infant to possessing a deepening of interests as an adult
- from having a short-time perspective as an infant to having a much longer time perspective as an adult

- from being in a subordinate position in the family and society as an infant to aspiring to occupy at least an equal and/or superordinate position relative to his/her peers
- from having a lack of awareness of the self as an infant to having an awareness of the control over the self as an adult

These characteristics are descriptive of a basic developmental process along which the growth of individuals in our culture may be measured.

In the work environment, because or our management principles, people are: (1) provided minimal control over their workday world; (2) expected to be passive, dependent, subordinate; (3) expected to have a short-time perspective; (4) induced to perfect and value the frequent use of a few superficial abilities; and (5) expected to produce under those conditions leading to psychological failure. All of these characteristics are incongruent, in conflict with those which healthy human beings desire. They are much more congruent with the needs of infants in our culture.

Argyris then suggests that the conflict leads to the following escapes for the workers:

- leaving the organization
- climbing the organizational ladder (promotion)
- manifesting defense mechanisms such as daydreaming, aggression, ambivalence, regression, projection, and so forth
- becoming apathetic and disinterested toward the organization, its make-up, and its goals; this leads to such phenomena as: (a) reducing the number and potency of the needs they expect to fulfill while at work, (b) goldbricking, setting rates, restricting quotas, making errors, cheating, slowing down, accidents, and so on
- creating informal groups to sanction the defense reactions and the apathy, interest, and lack of self-involvement
- formalizing the informal group
- evolving group norms that perpetuate the above behavior
- evolving a psychological "set" against the company

Then, because of the employees' actions, many managements tend to respond to this behavior by:

- increasing the degree of their pressure-oriented leadership
- increasing the degree of their use of management controls

- increasing the number of pseudo participation and communication programs

These three reactions by management actually compound the dependence and subordination that the employees experience; this compounding, in turn, causes the employees to increase their behavior, the very behavior management desired to curtail in the first place.

Argyris' incongruency theory has a great deal of meaning in safety. As employees adapt to frustration and conflict in their work situation and manifest defense reactions such as daydreaming and aggression, these surely influence the accident record. As employees exhibit the various forms of escape, our safety record is influenced to a very marked degree. When we see these things, our reaction is very much as Argyris indicated:

- We increase the degree of our pressure to enforce safety rules (three-day layoff).
- We increase the degree of our controls. We watch even closer-and step up our activities.
- We increase the number of pseudo participation and communication programs (posters, contests, gimmicks). This is real. We have all seen exactly these reactions toward work.

What should we do in safety programming? As Argyris suggests, we should consider programs that reduce the degree of dependency, subordination, and submissiveness. We should build safety programs that incorporate the concepts of job enrichment, participation, and employee-centered leadership.

Job Enrichment Theory

Although Chris Argyris and other behavioral scientists came to the conclusion that job enrichment is a necessity, it was actually Frederick Herzberg who first coined the term and expressed the principle. Herzberg's interest in job enrichment grew out of his discovery of what might be called his dual-factor theory of *job satisfaction and motivation.* Herzberg initially reviewed the literature (3,000 books and articles) in the field of job attitudes and motivation in his book *Job Attitudes: Review of Research and Opinion,* a scholarly review of what was known about attitudes from 1900 to 1955. In analyzing the data reviewed, he noticed that the things people said positively about their job experiences were not the opposite of what they said negatively about their job experiences. The reverse of the factors that seemed to make people happy in jobs did not make them unhappy. That was the core insight that led to Herzberg's theory. Job satisfaction and job dissatisfaction are not opposites; they are a completely separate continuum, like hearing and vision. If this is true, then they must be produced by different factors and have their own causes.

The essence of Herzberg's Motivation-Hygiene theory can be illustrated by comparing the traditional way of viewing dissatisfaction and motivation as simple opposites on a single continuum with the motivation-hygiene model that places motivation on a separate and distinct continuum from dissatisfaction. The classical approach to motivation, according to Herzberg, has concerned itself only with the environment in which the employees work, that is, the circumstances that surround them while they work and the things given in exchange for work. Herzberg considers this concern with the environment a never-ending necessity for management, but it is not sufficient in itself for effective motivation. That requires consideration of another set of factors, namely, experiences that are inherent in the work itself.

Dissatisfiers seem to be such items as pay, benefits, company policies and administration, behavior of supervisor, working conditions, and other factors that are generally peripheral to the task. Though traditionally thought of by management as motivators of people, these factors are actually more potent as dissatisfiers. High motivation does not result from their improvement. Dissatisfaction does result from their deterioration. Motivators are such items as achievement, recognition, responsibility, growth, advancement, and other matters associated with the self-actualization of the individual on the job. Job satisfaction and high production are associated with these motivators, while disappointments and ineffectiveness were usually associated with the dissatisfiers. A challenging job that allows a feeling of achievement, responsibility, growth, advancement, enjoyment of the work itself, and earned recognition motivates employees to work effectively. Factors that are peripheral to the job (work rules, lighting, coffee breaks, titles, seniority rights, wages, fringe benefits, and the like) dissatisfy workers. These could be considered tease items which will temporarily please but which will be long-range motivational failures. In other words, the factors in the work situation that motivate employees are different from the factors that dissatisfy employees. Motivation stems from the challenge of the job through such factors as achievement, responsibility, growth, advancement, the work itself, and earned recognition.

Why discuss these old theories that we all know? (1) Because they influence behavior mightily. They are behavior-based and researched-based. (2) Because, as old as they are, safety programs, in most cases, still ignore the concepts. We build rules and regulations, use JSAs and SOPs that do not allow employee thought (treat them immaturely) and we build safety programs primarily out of dissatisfiers.

Current management theory understands and has been striving for years to build cultures that treat people with maturity and respect, to provide the opportunity for achievement, recognition, responsibility, etc. The excellence in management books and books on culture all bear out the concepts from the motivation research. Until very recently, safety has done a rather poor job of using the concepts.

WHAT IS BEHAVIOR-BASED SAFETY?

Behavior-based safety is the building of a management system in safety that at each level of the organization defines precisely what behaviors are required, measures whether or not these behaviors are, in fact, present and reinforced regularly (daily, hourly) for the behaviors desired. This obviously is needed from the worker to the CEO, as behaviors are needed at each level.

Behavior-based safety is, in fact, one of the primary answers to future improvement, but only if and when it does the following:

- defines the behaviors needed at each level of the organization, from bottom to top
- ensures each person clearly understands the required behaviors
- measures whether the behaviors are, in fact, there
- rewards (reinforces) for the behaviors on a regular (daily, hourly) basis

Behavior-based safety systems that are real management systems do not turn safety over to any one level of the organization. It is great to get worker involvement—in fact, it is crucial. But to eliminate and isolate management from the process is dangerous to both sides. Management has the legal accountability for safety. To turn over the authority for safety, retaining the responsibility and accountability is tricky at best, dangerous at worst.

One of the fundamental tenets of safety is that safety systems can look like almost anything, as long as these systems meet the following two criteria:

1. There is a system of accountability in place that defines roles, ensures

knowledge to fulfill the roles at each and every level from worker to CEO, measures the role fulfillment; and has rewards contingent upon role fulfillment.

2. The system asks for, allows for, requires, and ensures participation at each level.

A true behavior-based approach to safety requires behavioral measurements. Upstream, it requires a statistically valid observation/sampling process (Are people behaving more safely today than last week/month?). To achieve statistical validity, a number of consistent observations are needed over time—one observer looking at all unsafe behaviors, not just a selected few.

Downstream, statistical validity is also needed. For most of us, that means shedding the old measures (frequency rates, incident rates, severity rates, etc.) and replacing them with valid, meaningful measures (process improvements achieved, perception surveys, etc.).

USING BEHAVIOR MODIFICATION

The behavioristic concepts are even less used today in safety and management. Trying to use these concepts is perhaps what the behavior-based proponents are really talking about.

Safe behavior reinforcement relies on establishing mostly positive consequences for performing work-related tasks in a safe manner. The normal approach to safety in the workplace is for management to apply an averse consequence (punishment) for unsafe acts. The ineffectiveness of punishment and disciplinary action in reducing unsafe practices is well known.

A second management approach is to rely upon avoidance conditioning. That is, after experiencing an accident as the natural consequence of unsafe behaviors, the employee will learn over time to avoid unsafe conditions or practices. However, since accidents are relatively rare events, resulting from one of perhaps thousands of unsafe behaviors, and since not all workers experience an accident, avoidance conditioning tends to have little effect in reducing unsafe behaviors.

In fact, unsafe behaviors produce certain rewards and thus may tend to be repeated. The simplest, most efficient, and most effective way to accomplish a task may not be the safest or the most rewarding way.

Unsafe practices generally result in a few punishments (reprimands or injuries) but may allow increased productivity or time for other activities, while safe practices earn little recognition from supervisors or co-workers and may involve a time cost. Also, even safe workers may be the victims of an accident caused by factors beyond their control (a design error). Thus the motivation to perform safely is generally insufficient in light of these competing factors.

Typically, the emphasis of most reward systems used in safety is on the negative, the reduction of accidents, with little guidance about what individual workers should do to avoid accidents. When desired performance is rewarded, the rewards are so infrequent as to be virtually ineffective.

Because natural consequences frequently do not reinforce safe behavior (and in fact may punish through reduced time or productivity), a change is required that will provide positive consequences for safe behaviors. These consequences must be more significant to the worker than the competing positive consequences of unsafe behaviors.

This shifts the focus from unsafe to safe behaviors and provides management with the opportunity to reward safe behaviors rather than to discipline unsafe behaviors. Supervisors must regularly and immediately reward safe acts.

The immediacy of reward is critical. Annual safety award presentations and performance appraisals are ineffective in improving plant safety. Likewise, consistency is important: supervisors who praise productivity probably are inadvertently praising unsafe practices or chance taking.

A behavioral safety program begins with identifying and defining desired behaviors, setting goals and standards and providing instruction. The consequences of safe behavior is a positive reward for safe acts.

The emphasis is on observable behavior that can be measured (that is, performed in either a safe or an unsafe manner).

The positive program involves providing employees with predetermined consequences for behavior. Safe practices are rewarded with tangible benefits, social reinforcement (such as praise) or information (feedback) about performance.

Subsequent maintenance of the desired safe behaviors depends on continued but less frequent reward than does establishment of these behaviors. The efficiency of the reward system in maintaining behavior depends on the frequency of reward (schedule of reinforcement).

Thus, regular reward is important in establishing a desired level of safe practice, but its maintenance can be achieved through less frequent and intermittent rewards. Should reward no longer be offered, however, the level of safety behaviors will return to its original level.

The overwhelming evidence collected from actual worksite investigations suggests that positive reinforcement programs are effective in increasing the level of safe performance and that this level may be maintained over the long-term. Unlike most training programs, a direct link can be demonstrated between the program and actual performance on the job.

By avoiding disciplinary or punitive measures, job satisfaction, morale and supervisor-employee relations are improved. As a result of the program, new standards for safe performance are also established within the work unit and their maintenance is reinforced by peer pressure.

The series of "One Minute Manager" books attempted to build behaviorism into management philosophy. The "One Minute Manager" concepts of a 4-to-1 relationship of positive to negative reinforcement, of finding people doing things right, that training may build behavior but reinforcement maintains it, are all behavioristic concepts accepted by management, but still not readily accepted or used in safety. Our experiments in over 50 studies show that positive reinforcement following safe behavior, if applied consistently, will reduce unsafe behavior by an average of 40% in a relatively short time period (see D. Petersen, *Safe Behavior Reinforcement,* Goslen, NY: Alorey, 1989).

Several years ago, the Association of American Railroads, funded by the Federal Railroad Administration, embarked upon an experiment to learn if the concept of positive reinforcement would be feasible and effective in an industry as different as railroading.

A project study team composed of safety directors of some of the major U.S. railroads and myself as an outside consultant was formed. In earlier years, this project study team had done major work in administering a survey to several pilot railroads to better understand the perception of workers, supervisors, and three levels of managers of railroad safety programs.

Phase II of this research effort was to determine if some new approach might impact those accidents on the job caused by human error. Traditionally railroad statistics show that a major portion of all accidents are caused by human error.

In each of two pilot railroads, a group of 75 to 150 first-line supervisors were selected to received experimental training in human behavior, accident causation, human error reduction, positive reinforcement techniques, and employee assessment techniques. The cooperation of their supervisors was also sought to assure that the techniques taught would be used by the supervisors once back on the job after the training.

An experimental design was conceived to establish whether or not the program was effective. On each railroad, one division was selected as an experimental division (the division to receive training and implement the reinforcement techniques), and an adjacent, similar division selected as a control division (to receive no training and to do nothing differently).

For each of the four divisions, as many measures of safety performance were taken as possible, both before and after, to best judge the results. Measures used were as follows:

- knowledge attained by supervisors as a result of the training experience, before and after tests were used
- reactions to the training and to the technique of positive reinforcement were asked, initially and after three months of using the technique
- performance of supervisors, before and after three months, was measured by asking each to report the number and kinds of activities engaged in the week prior to training, and the week prior to the three months follow-up training
- performance of the work force as measured by safety sampling, a statistically valid sample of worker behavior, sampled before training, after three months, and after six months

Samples were taken in all cases by safety professionals, and the same samplers were used throughout the entire study period to ensure accuracy and uniformity.

Accident results were not used as a measure of whether or not the program was effective. Whether or not accidents happened is simply an invalid measure of effectiveness due to the small group size, and the short length of the study. One division's record for six months is not a large enough sample to tell the facts needed to judge effectiveness of any program. Table 3-1 shows the results for these measures as written by the project committee in the executive summary presented to the top executives of each railroad by the consultant. The safety sampling results were the primary measure used to determine program effectiveness. As shown in Table 3-1, in the experimental division there was an improvement of performance of 40% in one railroad and 49% in the other. Changes in both control divisions were insignificant. All measures showed significant improvement in the experimental divisions compared to the control division counterparts.

		RRI	RRII
I. Knowledge—gain as measured by pre- and posttest scores	A. Average pretest score	4.54	3.58
	B. Average posttest score	23.57	28.50
	C. Gain	519%	769%
II. Reaction to training and use of techniques taught	A. Average of reaction to	4.38	4.23
	B. Positive reaction to being asked to use technique (scale 0-5)	4.29	4.18
	C. Positive reaction to using	4.04	4.05
III. Performance reported by supervisors	A. Reported increase in use of techniques taught as measured by what supervisors described themselves as doing before and after training	40%	14%

		% of Unsafe Behavior Observed			
		RRI	Control	RRII	Control
IV. Performance of work force as measured by sampling of worker behavior before and after training supervisors	A. Prior to training	35%	34%	39%	31%
	B. Three months after training	20%	34%	23%	29%
	C. Six months after training	21%	31%	20%	30%
Improvement in performance		40%	8%	49%	3%

V. Accident performance for the two areas was influenced by a number of variables not related to the study. A short-term comparison of reported accidents for the experimental and control groups on both properties shows "experimental groups" experiencing fewer injuries than did their respective "control groups."

Figure 3-1. Results of Positive Reinforcement Experiment.

As indicated in V in Table 3-1, the accident record, while not regarded as a judge of results, did improve considerably more in experimental than in control divisions. One railroad showed a 52% reduction in the accident record for the period of the study compared to the previous period.

It would seem that this training did, in fact, have an impact in both experimental divisions. The project study committee of the Association of American Railroads made the following statements on the basis of the results obtained:

> *1. It is possible to significantly improve safety performance through educating first-line supervisors to the human behavior problems associated with accidents, training them to apply positive reinforcement techniques in their daily activities and providing support through the organization for their efforts.*
>
> *2. It is possible to measure safety performance through observations of employee behavior. This may be a much more reliable technique than the use of accident statistics, in the short term, for determining the effectiveness of a company's safety program effort.*

An outline of the training provided supervisors in both experimental divisions is shown in Figure 3-2. Six hours of prior training was given with a three-hour follow-up session three months later. All training was by the consultant.

More important than this training, however, is the fact that supervisors were required to use some of the training when back on the job. In each case, the division manager was asked to devise a simple system of accountability: each supervisor had to report utilization of the concepts learned to his/her boss weekly. In each case, the supervisors were asked to observe workers as usual and to report to their boss the number of observations made each week.

Second, supervisors were asked to act following each observation. If the worker observed was working unsafely, the supervisor was asked to deal with the infraction as s/he normally would. If, however, the worker was working safely, the supervisor was required to act also; to contact the worker and positively reinforce that desired behavior—to do something with that worker that might make that worker more likely to work safely in the future. Each week each supervisor reported to his/her boss on a small 3 x 5 card the number of observations made, and the number of positive reinforcements given.

The experiment seems to indicate a positive value to the training and to the techniques used.

SESSION OUTLINE

Hours 1-2 Introduction

- Overview of the study
- Introduction to the session
- Pre-tests and questions

Why Try a Different Approach to Safety?

- Updating safety theory
- What we've always done versus what we now know
- What accidents are and what causes them
- The three Es of safety versus the three behavior changes
- What other companies and industries have found

Hour 3 Worker Motivation and What We know About It

- What motivates is
- A person doing whatever is necessary to satisfy his current needs
- What needs are
- The major theories and concepts
- Needs of most workers
- Needs change
- Motivators and dissatisfiers
- What the research shows

Hour 4 Influencing Worker Behavior

- Your three choices
- Motivation—the environment
- Attitude change
- Changing behavior
- Motivation—the easiest way
- What influences workers usually
- The peer group
- You and your style of leading
- Your credibility (particularly in safety)
- Your attitude toward safety (what they see)
- How you measure and judge them
- What they think your priorities are
- The organizational climate

Hour 5 What You Can Do

- Understand each worker
- Understand the group of workers you supervise
- Use the motivators

Hour 6 Using Positive Recognition

Figure 3-2. Supervisor training outline.

REFERENCES

Argyris C. *Personality and Organization.* Baton Rouge, LA: Hooper, 1957.

Blanchard K. *The One Minute Manager.* Los Angeles: Berkley, 1981.

Herzberg F. *Job Attitudes: Review of Research and Opinion.* Pittsburgh: Psychological Service of Pittsburgh, 1957.

Herzberg F. *Work and the Nature of Man.* Ada, MI: World, 1966.

4 Definitions

Although everyone agrees on the necessity of employee involvement, when we come around to doing it, we get very confused. One of our first problems is defining terminology. What do we mean by employee involvement? Do we mean we would like input from employees so that we can make better management decisions? Do we mean that we would like to share decision making with employees? Or do we mean we would like to turn decisions over to the employees? This chapter will briefly cover definitions of major terminology related to employee involvement. Parts II and III will explore in-depth the implications of putting into effect the various options for employee involvement.

Defining involvement is easy; getting it is difficult. But getting it starts with defining precisely what you and your organization mean by involvement and participation.

Our ultimate goal should be authentic involvement:

Real, authentic involvement is worker participation where the worker engages in activities that satisfy that person's human needs, and thus are motivational. Every person's needs are different—but most people respond to being treated like a mature, adult human being; being treated as an equal, not a subordinate; being able to use their innate intelligence and skills daily, even hourly; being able to achieve; being given responsibility; and being recognized for doing a good job.

This seems simplistic, but the decision to authentically involve workers is seldom made by an organization before embarking on an involvement course. Similarly, we tend to confuse, or not think about our terms. It is common to find organizations that talk in terms of self-directed teams, employee empowerment, and participation, as if they were all the same. They are not. Actually they could not be more different. The distinctions between these terms will be clarified in this chapter.

In safety journal articles, safety advertising, and speeches at safety meetings, there seems to be a pervasive feeling that behavioral approaches consist solely of involvement of workers in determining how they will most likely hurt themselves and in peer observation and intervention. Or, if you've attended any of the behavioral symposiums, you've seen there are two options: peer intervention or culture improvement. Are these the only choices we have to change behavior? Or are these simply the packages that are for sale?

I believe there are many other options available to industry (see Figure 4-1). These options fall into two major categories:

- options that involve a change in the organizational structure
- options that can be done within the current management structure

Involvement Options with Change of Organizational Structure

In attempting to get employee involvement, many management people start with the assumption that they must change the hierarchical structure in order to get the participation of the work force—an assumption that is sometimes fatal flaw number one. Going to self-directed work teams does require a change in the structure because there will no longer be supervisors. But simply removing supervisors does not get self-directed teams. Depending on your previous culture, simply removing supervisors might get some self-direction, but more likely will bring in chaos.

Self-Directed Work Teams

Some organizations keep the supervisors and simply change their title to team leader or facilitator, which only leaves the former supervisor with confusion as to his or her new role and leaves the workers with no leadership. Self-directed work teams can work very well as long as there is a logical transition, sufficient maturity in the work group, and a clear definition of what

INVOLVEMENT OPTIONS

Without Changing the Structure		
Input for Better Decision Making	Shared Decision Making	Limited Empowerment
Committees	TQM	Some Management Intervention (Instructional, Supportive, Motivational)
Safety Improvement Teams	SPC	
Ad Hoc Committees	Decision Making	
Task Forces	TOR	
Involve in Day-to-Day Operations (Meetings and Problem Solving)		
One-Minute Manager Concepts		
Objective-Setting Approaches		
Situational Leadership		
Peer Observation and Intervention		
Behavior-Based Safety		
Incident Recall		
Job Safety Analysis		
Hazard Hunts		
Ergonomic Analysis		

By Changing the Structure	
Total Empowerment	Self-Directed Work Teams

Figure 4-1. Involvement Options.

the team is responsible for and how they will be held accountable. I have seen teams do a superb job in safety (and other areas) in diverse industries, such as petroleum, railroading, steel, and food processing. But without clear definitions, I have seen attempts to change to self-directed work teams produce chaos in industries such as utilities, railroading, steel, food processing, and paper milling.

Before you remove all your supervisors, you might wish to determine your organizational readiness and the maturity level of all employees. Often neither workers nor midmanagement is ready. Many things determine readiness: the current culture, corporate traditions, the degree of trust that exists, management's credibility, etc. These things are measurable.

Also be sure your organization is not removing supervisors just to save money, whether they call it reengineering, right-sizing, revitalization or whatever. Involvement and downsizing are not the same. Downsizing (particularly through attrition) simply removes the most experienced (and best) people. It leaves you with less experience and less knowledge than before. With the removal of supervisors you have a number of unqualified people, unsupervised, running the operation. Is this good? You decide.

Going to self-directed work teams can be great, but it doesn't just happen. It usually takes some role redefinition, training, and personnel adjustment.This can be an excellent way to go as long as you have figured out ahead of time exactly how you will hold the teams accountable for specific performances, like safety. Some companies have thought this out and clearly specified exactly how the team will be measured and rewarded for safety performance.

Employee Empowerment

We also talk a lot about employee empowerment, at times even confusing it with teams. With true empowerment the individual worker, not the team, decides how he/she will work. The two approaches are almost opposites. True empowerment would allow the worker to decide whether to follow safety rules, wear personal protective equipment, etc. With empowerment, there must be very clear guidelines as to where people are empowered and where they are not.

Empowerment is a wonderful concept. The behavioral people tell us that empowerment provides the employee the opportunity to achieve, to accept responsibility, and to be rewarded for his/her achievements—all true theoretically, and in many cases in actuality.

Empowerment is perhaps the ideal that management strives for: that employees will "grow up" and accept the responsibility for their actions. In almost every company I have worked with, the first wish of management is for employees to accept their "responsibility." Usually, I have found it to be true that employees are perfectly willing to accept their "responsibility" when management has first demonstrated that they are fulfilling their "responsibility." Management's responsibility is made clear in the Occupational Safety and Health Administration standards.

Employee empowerment is in fact a wonderful concept. It is the ideal for the perfect company. But in the imperfect company, it poses some problems. Empowerment suggests an employee can decide how to work and how safely to work. Is it OK for a worker to choose to work unsafely? Not to wear PPE? Not to follow the company procedures? Empowerment suggests it is OK—unless some ground rules are established ahead of time.

Empowerment is great for self-directed, mature adults, who would follow the safe procedures and regulations regardless of management's pressure to "get the job done." Empowerment is not as good when values are not clear, priorities shift with the situation, and management is unclear about their real wishes. Empowerment leaves it up to the individual to decide what he or she will do in any situation.

Involvement

Employee involvement or participation is quite different from empowerment and self-directed work teams. With involvement, the individual worker is asked to help solve existing problems because he/she has the knowledge and experience in the job that management does not have, can never have. The organizational structure can take many forms: quality improvement teams, quality circles, ad hoc committees, task forces, etc. Anything that works, but perhaps safety committees are the least effective choice. These approaches tend to get all employees on the team of problem solving/decision making while not turning everything over to the work force.

Involvement teams can go as far as you wish: input, problem solving, decision making. The point is to decide ahead of time what you are trying to achieve: input, shared decisions or employee decisions, team-run operations or employee ownership, etc. Then set the guidelines to make it happen and take the steps to get there.

INVOLVEMENT OPTIONS WITHOUT CHANGE OF ORGANIZATIONAL STRUCTURE

Creating involvement options that do not require changing the organizational structure is the easiest, quickest route to involvement. It is also the logical interim step to teams or empowerment. Figure 4-1 shows a variety of ways to involve the employee. Most of the ways will give the employee authentic involvement, that is involvement in meaningful, fulfilling activities, not just grunt work the boss doesn't want to do. In authentic employee involvement, anyone can be involved.

The least authentic is usually the traditional safety committee because typically only a few can serve on the committee. The safety committee structure can create a three-class workplace: management, involved employees, and noninvolved employees. This may create additional problems because the noninvolved employees must do the work while the involved sit in meetings. Although safety committees are traditional and they may be required by law in the future, they are perhaps the least effective route to employee involvement.

Improvement Teams

Improvement teams are, in effect, ad hoc committees—people working together for a short period, solving a specific problem, and presenting their findings and solutions to management. Improvement teams can be ad hoc committees formed within a department because the boss has asked for help or task forces formed by upper management to solve a specific systemic problem. A team might consist of workers targeting areas like the following:

- ergonomic fixes
- identifying hazards
- identifying traps
- identifying systemic problems
- assessing culture

Total Quality Management and Statistical Process Control

In many organizations management has applied the concepts of total quality management (TQM) to safety. They have involved many people in a process improvement, such as continuous improvement of the safety process. With some minimal training in the concepts of statistical process control (SPC) as a background, TQM teams have done excellent work in organizations to improve the entire safety system.

Involvement in Day-to-Day Operations

When a supervisor asks his/her people to help make things better, the supervisor is involving employees in day-to-day operations. If interested employees help run the show (analyze what is happening, solve specific problems, conduct meetings, make inspections, unearth hazards, etc.), they have a greater sense of authentic involvement in their workplace.

In 25-plus years of consulting, that is, interviewing workers, I have never found people who were not ready and willing to help make things better. It is true some people don't want to be involved in extra work; but even these people have great ideas on improvement, and others want to be part of a group working on improvement.

Involvement through Good Management

Situational leadership concepts (like the one minute manager) suggest that a good manager uses different approaches with each employee, depending on individual personalities and needs. The manager gets involvement from those who want it and at least listens and gets input from those who don't want involvement. Good managers get high involvement from their people without any artificial structure to achieve it.

Setting Objectives

When a leader involves his or her team in setting the safety activity goals (what must be done, who will do it, when it will be completed, etc.), then there is authentic involvement. This approach requires no structural changes or elaborate programs—just good management with high participation and involvement.

BEHAVIOR-BASED SAFETY

Peer observation and intervention are always an option—if it fits the culture. In some organizations, it works well and employees are enthusiastic. In other organizations, it is perceived as "ratting on your brother." If the behavior-based safety approach is used, there must be prior planning and training so that employees know what to observe and how to effectively intervene. There must also be planned, ongoing feedback to measure the effectiveness of the behavior-based approach.

MEASUREMENT

How can you accomplish safety in a team-driven or employee-empowered organization? It is possible because many organizations have done it. It all comes back to accountability—do people know what they are supposed to do each day? Are they, in fact, measured to assure they are doing the defined tasks? Is there a reward structure for doing the tasks? Penalties for not doing the tasks? Some companies have done this. They have clearly defined what they want employees or teams to do to satisfy their safety responsibility. They measure to see if the employees or teams do carry out the defined tasks, and they reward for accomplishments. Other companies find this process too difficult, and thus they don't get world-class results.

We have known for years what it takes to get world-class safety performance. In 1967, the National Safety Council researched what it takes. The results of the research, published in the National Safety News, August 1967, were uncomfortable—they didn't give easy answers. The survey said that what management does every day determines the results you get, from first-line supervisor performance to CEO performance. In 1992, 25 years later, a follow-up study was done by the same researchers. The results showed almost no change: What management (from CEO to first-line supervisor) does every day dictates what the accident record will be.

I was recently asked to speak at a national convention on Safety and the Bottom Line. It seemed like a good title, vague enough that I could use a standard speech and bend it to the title. And then I sat down and simply couldn't figure out what to say.

Safety excellence and the bottom line is a no-brainer, right? It's almost axiomatic. Why then is it so hard to convince companies to achieve excellence in safety?

I went back to my old text books—to H. W. Heinrich, who showed the 4-to-1 relationship of hidden costs to direct costs; to Frank Bird, who increased that ratio with his "iceberg"; to John Grimaldi and Foster Rinefort, who showed how to quantify hidden costs. And I still didn't come up with enough money to get management attention. One company (a utility) I'm working with has claim costs of $500,000 per year. They spend $50 million per year on coal in their fossil plants; they also have nuclear and hydro plants. As David Terhune said in a recent behavioral symposium, Safety to most executives is "small potatoes"—it is not going to get their attention.

We cannot, will not get results in safety by citing costs; they aren't that big in most situations. But, to quote David Terhune again, the process of safety (not safety itself) is the real "jewel." We can't quantify our activities in safety to convince executives and line managers—the numbers just aren't there. But when we sell the "process of safety" (the establishment of accountabilities, the measurement of performances, the reward for performance), when we build a system of getting employee involvement around meaningful things, then we begin to establish some meaningful, measurable things.

Whatever options are chosen (and there could be different ones chosen in different locations or departments), there always needs to be some measurement to tell whether or not it is working, whether results are being achieved. You'll need valid upstream measurements to tell if the process is, in fact, continuously improving, and valid downstream measures to see the results.

The upstream measurements are the proactive objectives being met, such as the following:

- number of people involved
- number of objectives met
- number of peer observations
- number of hazards identified and removed
- number of traps identified and removed
- has culture been assessed and dealt with, etc.

An additional upstream measure is behavior sampling: are employees working more safely as a result of the processes installed? Sampling used as a measurement tool (not as an intervention) can give an excellent picture with statistical validity of whether the processes are getting results.

Depending on the size of the organization, you may have to choose a different downstream measure to assess progress. Measuring by the number of people hurt (or any subset of that) usually measures luck more than whether the system is improving. Perception surveys have a much better validity as a primary downstream measure.

REFERENCE

Planek T, Fearn K. Reevaluating occupational safety priorities, 1967 to 1992. *Professional Safety.* October, 1993.

II

Options without Change of Structure

5 True Empowerment versus Limited Empowerment

Employee empowerment is perhaps one of the most common and most misunderstood terms in the discussion of involvement. In this chapter, we will discuss what empowerment really is and why it's so powerful. Then, we'll discuss a limited empowerment, applicable in the real world.

TRUE EMPOWERMENT

True empowerment would exist in the most perfect of all worlds, if each employee had the maturity level, ethics, intelligence, and motivation to be empowered and if each supervisor/manager could accept the accountability for performance while delegating the authority down, had the maturity, intelligence, and motivation to allow it. Empowerment really means that each employee has the power to decide what to do,how much to do, and what procedures and rules to follow in safety, quality, and productivity.

In a perfect world, a perfect company, with perfect employees this sounds like, and is, perfection. In an imperfect world, with imperfect companies, employees, supervisors, and managers, it may not be perfection, or even acceptable. A truly empowered employee may well decide that working in the manner he/she chooses is okay—without personal protective equipment, proper guards on machines, regard for housekeeping, or adherence to production quotas constructed to coordinate with the next department. Such true employee empowerment can potentially lead to chaos.

True empowerment, then, may not be desirable—in fact, may not even be legal. Various Occupational Safety and Health Administration (OSHA) laws require that work, such as fork-lift truck driving, lock-out procedures, etc., be done in rather specific ways. Common sense suggests the same thing: that true empowerment implies that any behavior can potentially be acceptable. And that, obviously, is unacceptable. Later in this chapter, we'll discuss some concepts of empowerment limited by clearly defined guidelines.

Empowerment remains, however, a crucial concept in organizations. In safety, empowerment is the result of a management system that gives people what they most want and need: responsibility, the ability to achieve and thus be recognized, to be treated like mature, adult human beings, like an equal. A person who is treated in this way will feel empowered.

In the book/tape *Zapp! The Lightning of Empowerment,* the concept of empowerment is thoroughly explained through humor in the format of a play. To quote from the book:

> Zapp! The Lightning of Empowerment *turns today's most powerful business tool—the concept of empowerment—from academic theory into inspiring, invigorating* energy *that will enhance the success of any business.*
>
> *Zapp! It happens when a technician's nontechnical ideas are taken seriously, or when a manager invites help in solving departmental problems. It happens when people feel they are personally contributing to an organization's success, when they can influence how things get done, when their efforts are recognized and accomplishments rewarded. It energizes people to tackle daily challenges and to increase quality, customer satisfaction, and productivity.*
>
> *Zapp! The Lightening of Empowerment is a contemporary fable that shows what supervisors, managers, and organizations must do to create and maintain an empowered work force. The adventures of manager Joe Mode, as he seeks and finds the secret of empowerment, provide insights that anyone who works with other people can immediately apply in all business interactions.*
>
> *To succeed in today's competitive marketplace, companies must make an extraordinary commitment to quality and customer satisfaction. Zapp! shows how, by investing in employee growth, any company can create a truly empowered workplace where these goals become a reality.*

(Reprinted with permission from W. Byham, *Zapp! The Lightening of Empowerment,* Harmony, 1988. Copyright by Core Media.)

In *Zapp!*, the employees are empowered as their ideas are taken seriously by a boss who understands the need for listening, for responding and reacting to employees' ideas, for unleashing the power of the employees' brains.

LIMITED EMPOWERMENT

Whether an organization (or a person) is ready for empowerment can be defined by identifying how far the people in the organization have developed. E. Scott Geller has devised a model that can help assess an individual's (or an organization's) readiness for empowerment (see Figure 5-1). My simplified version of Geller's model is shown in Figure 5-2.

The simplified Geller model says the following:

1. If a person doesn't know how to work in the manner we want (that is safely), we must teach him/her.
2. If that person knows how to work safely but doesn't, we must reinforce the desired behaviors regularly.
3. If the person is behaving properly only because he/she is being forced, we must continue, even increase more often, the reinforcement (both positive and negative) as indicated.
4. If the person is behaving properly (safely) because he/she is beginning to believe it is the right thing to do, continue the reinforcement (intermittently).
5. If he/she behaves properly (safely) because it is now his/her habit, continue the intermittent reinforcement (mostly positive) forever.

Obviously, only when the work force is at the fifth stage listed above, will true employee empowerment fit. If the employees are not at that level, there must be some, perhaps many, limits on empowerment.

True empowerment can happen only when employees are at Scott Geller's unconscious competence level—when safe behavior is automatic behavior. If all employees are not at that level, some limits on empowerment will be necessary. We call this limited empowerment, or perhaps empowerment within specific guidelines.

Perhaps a sports analogy will best describe true empowerment versus limited empowerment. In football, we have two teams of eleven players competing. The goals are clear: get the football across the goal line more times than the other team (production of points). One team is totally empowered (but

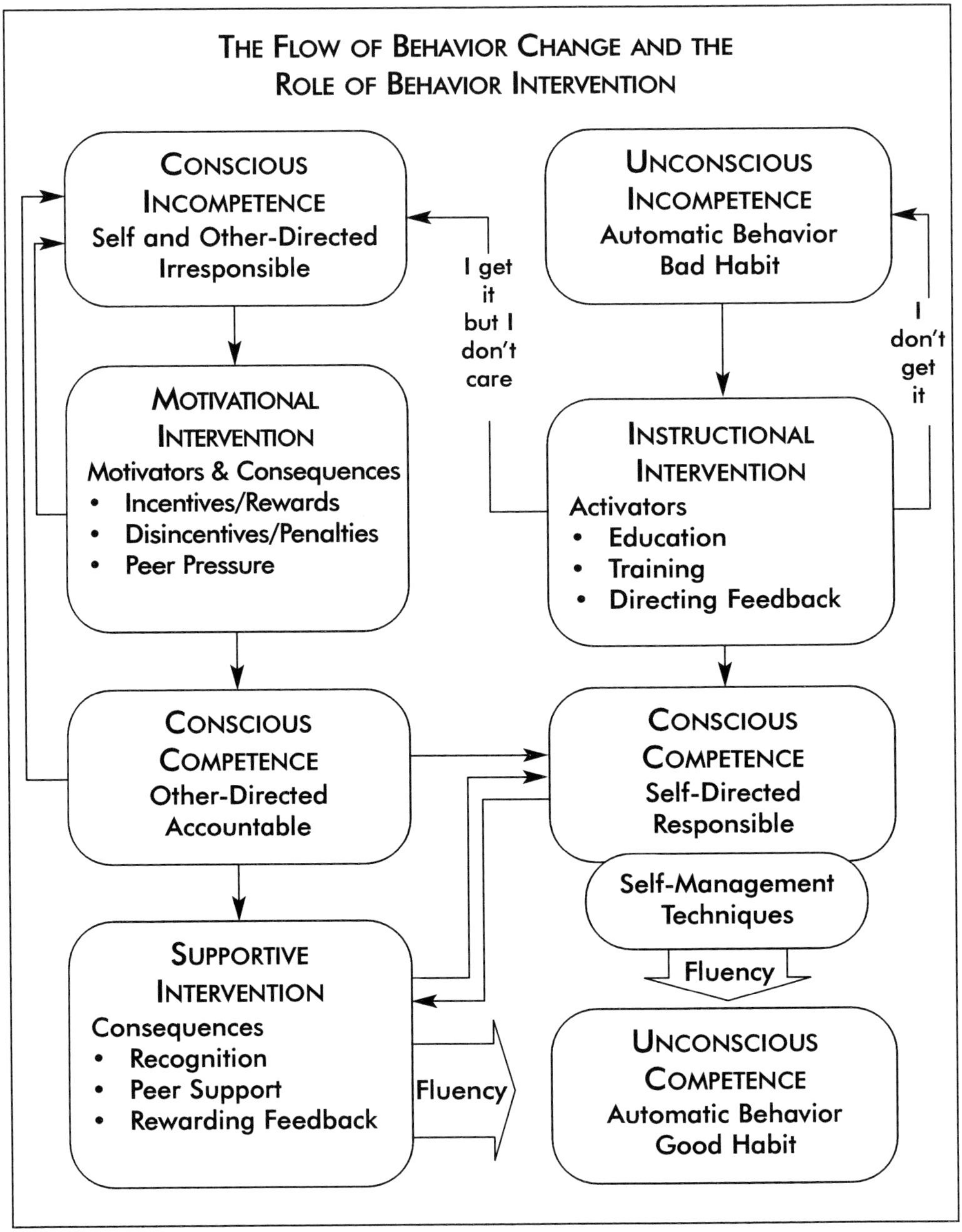

Figure 5-1. The Flow of Behavior Change and the Role of Behavior Intervention

(Reprinted with permission from E.S. Geller, Working Safe, *Radnor, PA: Chilton, 1996.)*

GELLER'S MODEL SIMPLIFIED

If They:	We Must:
1. Don't know	Educate/train
2. Know, but don't do	Use activators and consequences, with expectations, measures and reinforcement
3. Are doing because forced	Use support/reinforcement (consequences) often
4. Are doing it because they believe	Use reinforcement of consequences intermittently
5. Are doing it by habit	Use reinforcement intermittently

Figure 5-2. Geller's Model Simplified

not unconsciously competent). Not being unconsciously competent, they need some help to score. So they bring clubs, knives, machetes, and guns and use them at will (no safety rules). Also there are no sidelines, (no out of bounds), so they can run or pass anywhere and it's okay as long as they score.

The other team is equally empowered, with no rules, no regulations, no guidelines to the game. What we have created is not the game of football, but rather a war zone. Anything is okay as long as points (production) are scored.

Limited empowerment simply suggests that we all live within some ground rules. We all shall play with the same rules that will make the game fair and safe. Most football rules are, in fact, safety rules, and most teams find them easy to follow.

For most organizations, however, most employees are not at the unconscious competence level; therefore, they need some ground rules to follow. This is what limited empowerment means. It is first of all about what management decides the ground rules shall be. Is it okay to work without personal protective equipment? And if so, exactly when, where, and which equipment is it okay not to use? What and where are the sidelines to this ball game? What are the rules of behavior (the penalties) to this ball game? When does a player (or coach) get kicked off (or fired) from this ball game?

It becomes obvious that empowerment must be limited, and the question becomes how and how much. Geller provides some thoughts on this question in the following excerpt from his article "Behavior Based Safety: Confusion, Controversy and Clarification":

Three Types of Behavior

All voluntary behavior starts out as other-directed, in the sense that we follow someone else's instructions. Such direction can come from a training program, operations manual, or policy statement. After learning what to do, essentially by memorizing or internalizing the appropriate instructions, our behavior enters the self-directed stage. In other words, we talk to ourselves or formulate an image before performing a behavior in order to activate the right response. Often, we talk to ourselves after performing a behavior in order to reassure ourselves we performed correctly or to figure out ways to do better next time. At this point, we are usually open to corrective feedback if it is delivered well.

After performing some behaviors frequently and consistently over a period of time, they become automatic. A habit is formed. Some habits are good and some are not good, depending on their short- and long-term consequences. If implemented correctly, rewards, recognition, and other positive consequences can facilitate the transfer of behavior from the self-directed phase to the habit phase.

Of course our self-directed behavior is not always desirable. When we take a calculated risk, for example, we are choosing "willfully" to ignore a safety precaution or take a short cut in order to perform more efficiently or with more comfort or convenience. In this state, people are "consciously incompetent." Attempts to change self-directed behavior from incompetent to competent is often difficult, because usually it requires a relevant change in personal motivation.

Let's see what kinds of behavior-based interventions are appropriate for the three transitions referred to here:

- *Turning a risky habit (where the person is unconsciously incompetent) into self-directed behavior.*
- *Changing risky self-directed behavior (where the person is consciously incompetent) to safe self-directed behavior.*
- *Taking safe self-directed behavior (where the person is consciously competent) into a safe habit (unconscious competence).*

THREE KINDS OF INTERVENTION STRATEGIES

Instructional Intervention

Instructional interventions are typically activators or antecedent events used to move behavior from the automatic (habit) stage to the self-directed stage, or to improve behavior already in the self-directed stage. The aim is to get the performer's attention and instruct him or her to transition from unconscious incompetence to conscious competence. You assume the person wants to improve, so external motivation is not needed—only external and extrinsic direction.

This type of intervention consists primarily of activators, as exemplified by education sessions, training exercises, and directive feedback. Because your purpose is to instruct, the intervention comes before the target behavior and focuses on helping the performer internalize your instructions. As we've all experienced, this type of intervention is more effective when the instructions are specific and given one-on-one.

Supportive Intervention

Once a person learns the right way to do something, practice is important so the behavior becomes part of a natural routine. Continued practice leads to fluency and, in many cases, to automatic or habitual behavior. This is an especially desirable state for safety-related behavior. But practice does not come easily and benefits greatly from supportive intervention. We need support to reassure us we are doing the right thing and to encourage us to keep going.

While instructional intervention consists primarily of activators, supportive intervention focuses on the application of positive consequences. Thus, when we give people rewarding feedback or recognition for particular safe behavior, we are showing our appreciation for their efforts and increasing the likelihood they will perform the behavior again. And each consequence of the desired behavior facilitates fluency and helps build a good habit.

So after people know what to do, they need to perform the behavior many times before it can become a habit. Therefore, the positive reinforcement we give people for their safety-related behavior can go a long way toward facilitating fluency and a transition to the automatic or habit stage. Such supportive intervention is often most powerful when it comes from one's peers—as in peer support.

Motivational Intervention

Note that supportive intervention is typically not preceded by a specific behavior. In other words, when you support self-directed behavior, you don't need to provide an instructional antecedent. The person knows what to do. And, you don't need to activate desired behavior with a promise (an incentive) or a threat (a disincentive). The person is already motivated to do the right thing.

When people know what to do and don't do it, however, a motivational intervention is needed. In other words, when people are consciously incompetent about safety-related behavior, they require some external encouragement or pressure to change. Instruction alone is obviously insufficient because they are knowingly doing the wrong thing. In safety, we refer to this as taking a calculated risk.

We usually perform calculated risks because we perceive the positive consequences of the at-risk behavior to be more powerful than the negative consequences. This is because the positive consequences of comfort, convenience, and efficiency are immediate and certain, while the negative consequence of such behavior (an injury) is improbable and seems remote. Furthermore, the safe alternative is relatively inconvenient, uncomfortable, or inefficient, and these negative consequences are immediate and certain. As a result, we often need to add both activators and consequences to the situation in order to move people from conscious incompetence to conscious competence.

Motivational intervention is clearly the most challenging, requiring enough external influence to get the target behavior started without triggering a desire to assert personal freedom. Remember, the objective is to motivate a transition from conscious incompetence to a self-directed state of conscious competence. Powerful external consequences might improve behavior only temporarily, as long as the behavioral intervention is in place. Hence, the individual is consciously competent, but the excessive outside control makes the behavior entirely other-directed. Excessive control on the outside of people can limit the amount of control or self-direction they develop on the inside.

How can we facilitate a transition from the state of conscious competence/other-directed to conscious competence/self-directed? This is probably the most important and hotly debated question in the domain of behavior change. I have addressed this challenge in a recent book (Geller, 1998a).

Obviously, I cannot summarize a 250 page book here, but it might be useful to identify the topics of this book's five chapters. Each is relevant for developing a safety accountability system that promotes personal responsibility.

1. *Decrease top-down control for safety by (a) focusing on fact finding rather than fault finding, (b) discussing disadvantages of punishment, (c) considering differences between human error and a calculated risk, and (d) recognizing the role of interpersonal and environmental context on at-risk behavior.*
2. *Increase feelings of empowerment by (a) holding people accountable for numbers they can control, (b) setting goals that are specific, motivational, achievable, recordable, and trackable, (c) recognizing successive approximations toward major accomplishments and (d) increasing the use of supportive interventions, including behavior-based feedback, recognition and reward.*
3. *Help people feel important by (a) increasing opportunities for choice, (b) teaching principles and guiding customization of procedures, (c) demonstrating the significance of pro-active, behavior-based safety and (d) teaching the value of "emotional intelligence" (Goleman, 1995), including techniques for communicating more effectively with others and with oneself.*
4. *Cultivate belonging and interpersonal trust by (a) improving interpersonal communication, (b) building group consensus for important decisions, (c) promoting systems thinking and interdependence and (d) teaching and demonstrating the principle of reciprocity—the fact that helping others activates an obligation to return the favor.*
5. *Teach and support safety self-management by following the behavior management guidelines I present elsewhere.*

(Reprinted with permission from E. S. Geller, "Behavior Based Safety: Confusion, Controversy and Clarification," *Occupational Health and Safety,* January 1999, pp. 40-49.)

REFERENCES

Byham WC. *Zapp! The Lightning of Empowerment: How to Improve Productivity, Quality, and Employee Satisfaction.* New York: Harmony, 1988, quotation from dust cover.

Geller S. Behavior based safety: Confusion, controversy and clarification. *Occupational Health and Safety.* January, 1999, pp. 40-49.

Goleman D. *Emotional Intelligence.* New York: Bantam Books, 1995.

6 Shared Decision Making

If true involvement requires an unrealistic level of perfection and lmited involvement is not yet practical, the next option is to create an environment where employees can at least share in the decision-making and problem-solving process. In this environment, it is not management or labor, but rather management and labor sharing the process of problem solving, decision making, or both. This is a step between employee empowerment and traditional management, and in many cases a comfortable step, perhaps even an interim step to more complete involvement and empowerment.

In shared decision making and problem solving, both management and the work force can work together to solve the various problems that an organization faces. There is a recognition that the problems are bigger than either side can solve alone; both need to work together to make progress.

A number of things must happen for shared decision making and problem solving to occur. First, management must recognize that they simply cannot do it alone. The work force must also recognize that they cannot get to where they want to be without some cooperation from management. Neither can do it alone. Third, both sides must recognize that there must be a balance, perhaps even an equality, between the two. Neither side is in charge; rather, they are in it together and totally dependent upon each other.

Finally, having agreed on these points, both management and the work force must take the steps necessary to make shared problem solving and decision making really happen. This means both sides must be trained to be able to work together in their shared responsibilities. The training must include the following:

1. They must be trained in the process of problem solving and decision making.
2. They must be trained in how to work together in the processes learned.
3. They may need to be trained in how to get along with each other.

Training in the process of problem solving and decision making might include learning how to use specific analysis techniques, such as statistical process control (SPC) and the technique of operations review (TOR), discussed in the next sections of this chapter.

STATISTICAL PROCESS CONTROL

Statistical process control is simply the use of statistical concepts to control activities and improve their effectiveness. The use of statistical methods is not an all-purpose remedy for every corporate problem. But it is a rational, logical, and organized way to create a system that can assure continuing, ongoing improvements in quality and productivity simultaneously, and it can be applied to safety as well as production processes.

Although U.S. industry basically spurned the whole idea of SPC in the 1950s, now it is a popular concept wholeheartedly endorsed. How powerful a tool is SPC? When the United States ignored it, they found their productivity dropped 15% between 1960 and 1980. The Japanese bought it and increased their productivity by 150% in the same period. And their thrusts were for quality, not productivity. In the 1980s, American managers rushed to Japan to find their secret—only to find their secret was what the United States had rejected twenty years earlier.

The SPC method is really a two-pronged approach: using some SPC tools helps to solve problems and increases employee involvement. SPC tools that can be used most easily in safety by supervisors and others are the fishbone diagram, the Pareto chart, the flowchart, the control chart, and scatter diagrams.

A fishbone diagram was used in one company to brainstorm the various interrelated causes of a problem (Figure 6-1). It shows all possible contributing causes of the specific problem.

Pareto charts can be used in many ways in safety to plot types of injuries that have occurred in a department. The Pareto chart in Figure 6-2 was developed

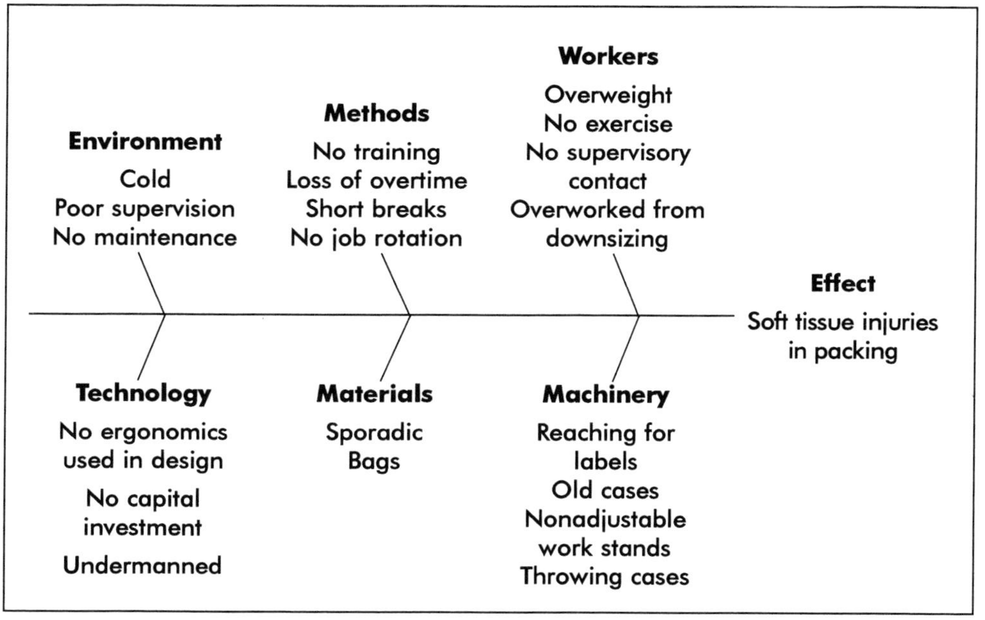

Figure 6-1. Fishbone diagram showing interrelated causes of a problem.

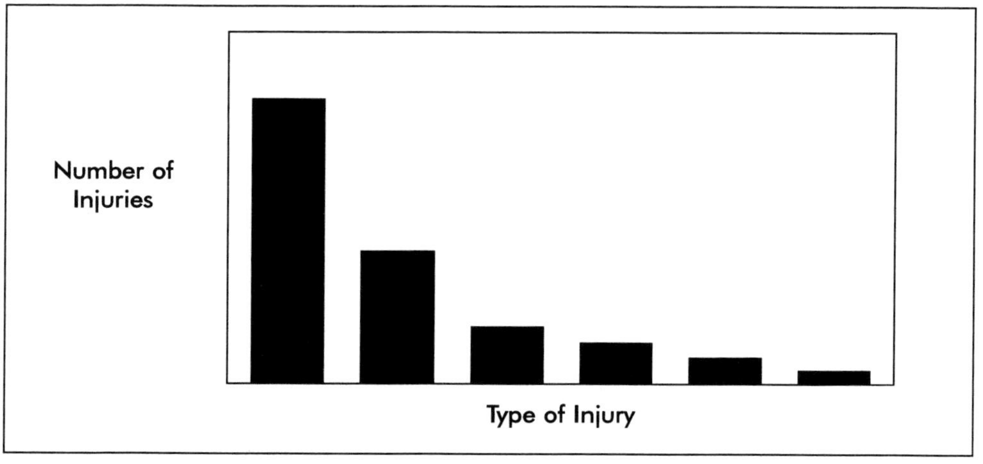

Figure 6-2. Pareto chart of types of injuries.

by a group of workers to identify the combinations of reasons they developed soft tissue injuries. The problem was the soft tissue injury in wrists, shoulders, and back. From Pareto charts, it was decided that a change in procedures could

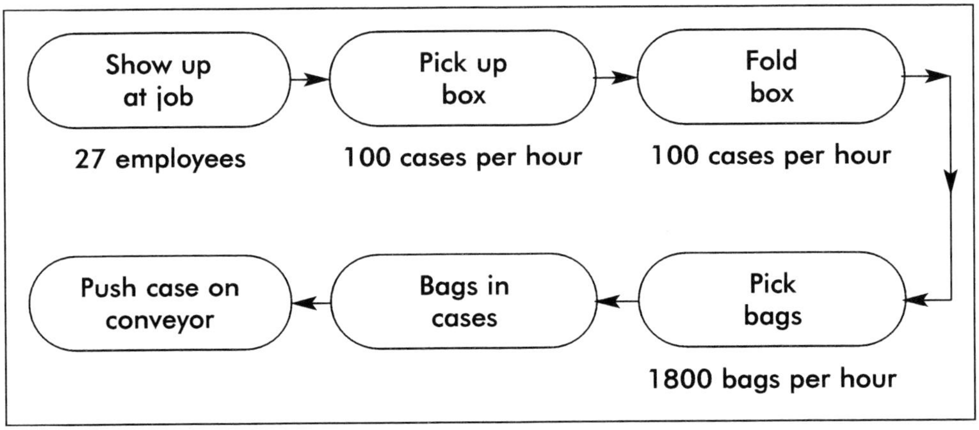

Figure 6-3. Flowchart of a process.

be made to reduce the soft tissue injuries in two departments: packers in the packaging department and box handlers in shipping.

The group next used a third SPC tool, a flowchart—basically a picture of the process, step by step (Figure 6-3). Finally, the group decided which factors to concentrate on:

- get adjustable packing tables
- do preshift warm up exercises
- use wrist bands
- rotate packers hourly

A substantial improvement was achieved. The approach became a normal way of life in the plant and was used on other safety, quality, and productivity problems.

By using SPC for safety issues, specific problems can be attacked and solved, and the system can be monitored. In using SPC, objectives can be set in terms of number of times the specific tools have been used, whether Pareto charts, fishbone diagrams, etc. More sophisticated measures might be percentage of accidents when fishbone diagrams were used; number of problems solved by employee groups using Paretos, flowcharts, etc.

The use of statistical safety control is one method of including employees into problem solving and decision making. Another is through the use of technique of operations review.

TECHNIQUE OF OPERATIONS REVIEW

The technique of operations review (TOR) was devised to assist you in finding some of the multiple, interrelated causes behind the accidents you are required to investigate. It is basically a tracing system. It also can be used as a training technique in safety.

In a tracing system, the investigator (you) is initially asked to identify what appears to be a major cause or factor behind an event (the accident). You select this cause from a long list of possible causes. Following the description of each major cause are numbers that lead you to other connecting factors, which might also be contributing causes.

One of the best of the tracing systems is the TOR, devised in 1967 by D.A. Weaver. The TOR centers around the Cause Code shown in Figure 6-4. The supervisor's incident investigation report in Figure 6-5 was developed for use with the TOR by Paul Mueller, then of the Green Giant Company.

The TOR begins with an incident. Its purpose is to expose what the real problems are behind this incident, which is viewed as a symptom of a more serious problem. The steps for using a TOR include the following:

1. Describe the incident. State clearly what happened.
2. Select one number from the Cause Code which seems to be the immediate cause of the incident.
3. Trace and eliminate. The initially chosen number is jotted on the form, and the trace step begins. Following this initial number and its description will be other numbers. Jot them down and read their descriptions. Then decide whether or not they could also be contributing causes to the incident. These numbers lead to additional numbers. List them, read the descriptions of these, and decide again. Keep tracing and eliminating numbers until they run out. When the "outs" overtake the "ins" or the final number on the list repeats the number at the top, you have come full circle. The numbers and descriptions that remain are those you have decided are contributing factors.
4. List the contributing factors.
5. Select solutions.

TOR Analysis—Technique of Operations Review: A hands-on process for managing operational error on the job. Originally developed for Wausau Insurance Companies.

1 TRAINING

10 Training not formulated or need not foreseen 23, 48, 64

11 Instruction was given but results show it didn't take 44, 47, 56, 75

12 Training available but the employee was not assigned or did not attend. 26, 35, 87

13 Performance not in accord with policy or procedure 47, 55, 62

14 Failure to provide training whose need had been specified 34, 83, 88

15 Error blamed on faulty training when in fact the error stemmed from deficiencies in management systems. 26, 36, 52, 81

3 DECISION & DIRECTION

30 By-passing, conflicting orders, too many bosses 33, 48, 80

31 Decision too far above the problem 34, 83

32 Authority inadequate to cope with the situation. 22, 82

33 Decision exceeded authority 13, 47, 86

34 Decision evaded; power to decide not exercised 25, 85

35 Orders or directives failed to produce desired action. Not clear, not understood, or not followed . . 41, 50, 52

36 Failure to investigate, and to apply the lessons of similar mishaps . . . 26, 43, 61

37 Hazard or problem—controls not developed 26, 64, 66, 86

TRACE	10	11	12	13	14	15		30	31	32	33	34	35	36	37	
GUIDE	20	21	22	23	24	25	26	40	41	42	43	44	45	46	47	48

2 RESPONSIBILITY

20 Duties and tasks not clear, or not accepted 22, 25, 40

21 Conflicting goals 30, 48, 83

22 Dual or overlapping responsibility 25, 30, 48, 80

23 Pressure of immediate tasks obscures full scope of responsibilities. 10, 32, 34, 87

24 Buck passing, responsibility not tied down 25, 48, 82

25 Job descriptions inadequate . . 48, 80, 84

26 Hazard or problem— not recognized 34, 37, 48, 81

4 SUPERVISION

40 Failure to orient or coach—new worker, unusual situation, unfamiliar equipment or process, etc. 23, 24

41 Information provided was insufficient to do the job right 23, 35, 48

42 Lack of two-way communication; failure to listen 15, 36, 82

43 Failure to correct behavior or procedure before accident/incident occurred 26, 36, 61, 73

44 Failure to supervise closely until proficiency was assured . . 23, 36, 48, 64

45 Honest error. Failure to act, or action turned out to be wrong 14, 15, 20

46 Disorder or confusion in work area 34, 54, 56, 87

47 Job practice out of step with job training 15, 42, 63

48 Initiative. Failure to see problems and exert an influence on them 23, 42, 50, 85

TOR Analysis Worksheets and materials
Licensing arrangements
Management and Employee Training

Available from R.G. Hallock PhD
Training Consultants to Business and Industry
Business and Technology Center Ste. 211
Pueblo, Colorado 81003
Phone (719) 584-2435

Figure 6-4. (Part 1) TOR Cause Code.

5 WORK GROUPS 5

50 Morale. Conflict, insecurity. Lack of faith in the leadership or the future of the job 31, 56, 66, 86

51 Lax leadership in example and attitude 31, 42, 74, 86

52 Team spirit. Failure to pull together, uncooperative 15, 21, 74, 81

53 Rules. Not publicized, not clear. Unfair enforcement or weak discipline 40, 41, 80

54 Clutter. Anything not needed in the work area 20, 32, 60

55 Lack of things needed—tools, space, protective equipment, storage bins, etc. 24, 36, 65

56 Work group sees little advantage to themselves in doing it right . . 21, 82, 84

7 PERSONAL TRAITS

70 Work assignment—unsuited for this particular individual 11, 33, 45, 63

71 Poor work habits; careless of rules, tools, equipment, procedures, etc. 26, 30, 53, 85

72 Health problem 26, 34, 80, 81

73 Inappropriate behavior or judgement 33, 45, 72, 83

74 Undesirable peer pressures influence work performance and risk taking 20, 56

75 Behavior not adjusted to the workplace 32, 53, 85

50	51	52	53	54	55	56	70	71	72	73	74	75			
60	61	62	63	64	65	66	80	81	82	83	84	85	86	87	88

6 CONTROL

60 Work flow. Inefficient or hazardous. Layout, scheduling, stacking, piling, routing, storing . . . 32, 46, 48, 66

61 Unsafe condition 36, 43, 65, 86

62 Equipment. Insufficient, unavailable, deficient design, inoperative. . 26, 31, 84

63 Procedure out of step with available technology; inadequate review and revision 24, 81, 85

64 Procedure not available or not followed 34, 80, 88

65 Deficient inspection, reporting, or maintenance. 34, 47, 88

66 Hazard or problem—controls not maintained 26, 34, 88

8 MANAGEMENT

80 Failure to assert policy and a management will before the mishap at hand. 15, 35

81 Goals are not clear or are not converted into decisions or directives. 21, 35, 85

82 Excessive span of control 30, 36, 85

83 Conflicting priorities not resolved. Excessive emphasis on short range accomplishments 12, 21

84 Departments inadvertently create problems for each other; inadequate coordination 34, 36

85 Inadequate development of subordinates; failure to encourage subordinates to exercise their power to decide 31, 50, 82

86 Inadequate appraisal and measurement of key goals and objectives . . 24, 37, 50

87 Insufficient staff to cover necessary functions; failure to use available human resources or to cope with turnover and absenteeism. 23, 46, 62

88 Hazard or problem—not properly evaluated 37, 55, 81

9 Brooks Place
Pueblo, Colorado 81001
(719) 542-3032
E-mail: weaveriz@home.com

Figure 6-4. (Part 2) TOR Cause Code.

INCIDENT INVESTIGATION

Employee (if involved) ______ Dept. ______ Clock No. ______

Incident date ______ Reported ______

1. Describe the incident. Include location, witnesses, and circumstances surrounding incident. Try to identify the casual factors involved.

2. Subject causes to TOR analysis: state, trace, eliminate.

3. List factors for which you will initiate corrective action.

   Factor: ______ Action: ______

4. List factors that require feasible corrective action by others. Circle routing to their attention.

Figure 6-5. Supervisor's incident investigation report.

The TOR helps to identify and define problems by searching for root causes. Such causes, if undetected, might lead eventually to accidents and other kinds of management losses.

TOR and SPC are merely examples of tools and techniques that can be used to share decision making. The important thing is to involve people at all levels in the process of decision making and implementation.

REFERENCE

Weaver D. *TOR—Technique of Operation Review.* Pueblo, CO: Weaver and Associates, 1967.

7 Input for Better Decisions

If an organization is not ready for shared decision making, another option is to create an environment in which input (ideas of workers) is asked for and considered before management makes their decisions.

This can take many forms. A informal relationship can be established between management and the work force in which sharing of ideas just happens and management's decisions clearly reflect the thoughts of the work force. A more formal relationship can be created in which management agreeds to consider workers' ideas before deciding what they will do about a particular subject.

In the input-of-ideas environment, management remains firmly in control of decision making and no attempt is made to allow the workers into the process. They may give input, may even help to solve problems, but it remains clear that decisions, rules, procedures, etc. are in management's hands solely.

The following sections explore some techniques to get employee input beyond just asking them for their help.

Safety Improvement Teams

Employee participation and involvement is one of the best ways to provide an environment that is motivational. When people have a piece of the action and are involved, it positively affects their behavior.

The safety-improvement-team approach involves calling employees together and telling them that you need their help in safety. Ask them to join with other workers to help solve problems in the department. Allow them to work with people of their choice and select problems of their choice. Provide them with any technical help they request. Ask for a regular (monthly) progress report.

JOB SAFETY ANALYSIS

Job safety analysis (JSA) is a procedure that identifies the hazards associated with each step of a job, and develops solutions for each hazard that either eliminates it or controls it. A job safety analysis worksheet is illustrated in Figure 7-1. In the left column of the worksheet, the basic steps of the job are listed in the order in which they are performed. The middle column describes how to perform each job step. The right column gives the safety procedures that should be followed to guard against hazards. The basic steps in the JSA process include the following:

- Select the job to be analyzed.
- Break the job down into successive steps.
- Identify the hazards and potential accidents.
- Develop ways to eliminate the hazards listed.

A blank worksheet used for Occupational Safety and Health Administration (OSHA) compliance is shown in Figure 7-2.

Selecting the Job

In selecting the job to be analyzed and establishing the order of job priorities, the National Safety Council suggests that you be guided by the following factors:

- Frequency of accidents. A job that has a repeated number of accidents is a high priority for a JSA.
- Severity of accidents or injuries. Any job that has produced disabling injuries might be considered a top priority for a JSA.
- A high potential for severity. If the potential for a serious accident or injury is present, a JSA might well be warranted.
- New jobs or changed jobs.

First, the job is broken down into its basic steps. These steps should describe what is being done in order of occurrence. The National Safety Council suggests the following key points in breaking down a job:

1. Select the right person to observe. The person selected must be experienced, capable, cooperative, and willing to share his/her thoughts.
2. Brief him/her on your purpose. If the worker selected has never worked

JOB: USING A PRESSURIZED WATER FIRE EXTINGUISHER		
What to Do *(Steps in sequence)*	**How to Do It** *(Instructions)* *(Reverse hands for left-handed operator.)*	**Key Points** *(Items to be emphasized. Safety is always a key point.)*
1. Remove extinguisher from wall bracket.	1. Left hand on bottom lip, fingers curled around lip, palm up. Right hand on carrying handle palm down, fingers around carrying handle only.	1. Check air pressure to make certain extinguisher is charged. Stand close to extinguisher, pull straight out. Have firm grip, to prevent dropping extinguisher on feet. Lower, and as you do remove left hand from lip.
2. Carry to fire.	2. Carry in right hand, upright position.	2. Extinguisher should hang down alongside leg. (This makes it easy to carry and reduces possibility of strain.)
3. Remove pin.	3. Set extinguisher down in upright position. Place left hand on top of extinguisher; pull out pin with right hand.	3. Hold extinguisher steady with left hand. Do not exert pressure on discharge lever as you remove pin.
4. Squeeze discharge lever.	4. Place right hand over carrying handle with fingers curled around operating lever handle while grasping discharge hose near nozzle with left hand.	4. Have firm grip on handle to steady extinguisher.
5. Apply water stream to fire.	5. Direct water stream at base of fire.	5. Work from side to side or around fire. After extinguishing flames, play water on smoldering or glowing surfaces.
6. Return extinguisher. Report use.		

Figure 7-1. Job analysis worksheet.

Source: Accident Prevention Manual for Business & Industry: Administration & Programs. *12th Edition Itasca, IL: National Safety Council, 2001, p. 177. Reprinted with permission.*

on a JSA before, thoroughly explain it. Ask the person for his/her help.

3. Observe the job for the breakdown
4. Record each step on the worksheet
5. Check the breakdown with the worker when you are done and get his/her input.

Job operation ______________________________

Presently required personal protective equipment ______________________________

Sequence of job steps	Hazards or OSHA violations
1.	1.
2.	2.
3.	3.
4.	4.

Recommended safe procedure

1.
2.
3.
4.

Figure 7-2. OSHA job safety analysis worksheet.

(Reprinted with permission from The OSHA Compliance Manual *by Dan Petersen. Copyright 1975 by McGraw Hill. Used with permission from McGraw Hill Book Company.)*

Identifying Hazards

After the breakdown, each step should be analyzed in detail to identify hazards and potential accidents. Each hazard or potential accident should be recorded on the worksheet in the center column, keeping it parallel with its step of the job procedure. Check with the employee you're working with for his/her ideas. Check also with other employees who have knowledge of that job.

When the hazards have been identified, the next step is to begin to develop solutions to the identified problems. Solutions might include the following:

- an entirely different way to do the job
- a change in physical conditions, layout, or environment
- a changed job procedure
- a change in frequency of how often the job is performed

For each hazard on the sheet, ask "What can be done differently, and how should it be done?" Answers and solutions should be very specific and very concrete to be of value. Solutions that merely state "Be more alert" or "Use more caution" or something similar are worthless. Solutions should state exactly what to do and how to do it.

To: ______________________________

From: ______________________________

HAZARD HUNT

I think the following is a hazard: ______________________________

____

____

DO NOT WRITE BELOW HERE—TO BE FILLED IN BY SUPERVISOR.

Supervisor:

Agree this is a hazard.

Corrected by Supervisor on ____________ Discussed on ____________

If can't correct, sent to Personnel on ____________________

Job order on _______ Scheduled _________Discussed on ____________

Do not agree this is a hazard.

Discussed ____________________Conclusion ____________________

To Personnel _________________ Conclusion ____________________

DO NOT WRITE BELOW HERE—FOR PERSONNEL USE.

Supervisor ______________________________ HH # ______________

Matrix # ______________________________ (Seriousness)

Figure 7-3. A hazard hunt form.

While performing the JSA, you'll learn more about the job observed than ever before. You also will have involved an employee and demonstrated to him/her and the department that you care about his/her safety on that job. And you'll be creating safer conditions for the job observed.

Hazard Hunt

Another method used successfully to spot possible accident causes and involve employees in the safety process is the hazard hunt (HH). The procedure can be implemented using the following steps:

1. Make copies of the hazard hunt form (Figure 7-3) available to all of your people.
2. Hold a short session with them to explain the form and the reasons for using it.

Degree of Hazard	Difficult Lengthy Costly	Moderate	Easy Fast
Major	4	2	1
	7	5	3
Minor	9	8	6

Figure 7-4. Safety matrix.

3. Have employees jot down anything they feel is a hazard and return the form to you.
4. Review the forms, correct those hazards you can, and initiate action when correction is out of your control.
5. Inform employees of your actions. Always tell them what you are doing, even if you decide the hazards they mention are not really problems. Then discuss any disagreements with individuals to clear the air.
6. If you agree something is a hazard and must be corrected, assign a priority to it and schedule it for rectification. The matrix in Figure 7-4 provides a good guide for prioritizing.

Ergonomic Analysis

An ergonomic analysis (EA) is a way of reducing the probability that employees will be injured through exposure to cumulative trauma or repetitive motion. Each job should be analyzed.

The primary focus of ergonomic analysis to prevent repetitive motion injuries, which are commonly called cumulative trauma disorders (CTDs). These CTDs are occupational injuries that develop over time, affecting the musculoskeletal and peripheral nervous systems. They can develop in any part of the body, but are most prevalent in the arms and back. These injuries are caused by jobs that require repeated exertions and movements near the limits of the individual's strength and range-of-motion capability. These movements, although not initially painful, cause microtraumas to the soft tissues. Over time, small strains to the muscle/tendon/ligament system build up, resulting in fatigue and soreness. If the individual continues the action that is causing pain, cumulative trauma disorders are likely to develop. CTDs can have the following effects on individuals:

- pain
- numbness or loss of sensation
- reduced strength
- degraded ability to perform work
- degraded ability to participate in leisure activities

Here are several good reasons why we should try to eliminate CTDs from the workplace:

- It's the right thing to do.
- It's the humane thing to do.
- CTDs are costly.
- OSHA will make you.

In order to prevent CTDs, we need to understand what causes them. Cumulative trauma disorders have been correlated with hazardous combinations of the following generic risk factors:

- forceful exertions
- frequent exertions
- body posture
- mechanical stress
- vibration
- low temperatures

The presence of these risk factors is related to the design of the job. For example, we can either design a workstation that requires the worker to use a poor posture all day or we can design it so that all operators can do the operation in a good posture. Areas of design that may expose the worker to the generic risk factors are called "job specific" and include items on the following list:

- workplace layout
- tools

Hand & Wrist	Possible CTD Problem
GRASP	
• pinch grip	• prolonged pinch grip
• static hold	• forceful grasp
WRIST POSTURE	
• flexion/extension	• flexion/extension>45°
• ulnar/radial deviation	• radial/ulnar deviation
FREQUENCY	
• hand or wrist manipulations	• > 10 per minute
MECHANICAL STRESS	
• localized pressure to palm or fingers	• prolonged exposure • intense exposure
• scraping/pumping	
• strike with hand	
• single finger trigger	
VIBRATION	
• high-frequency vibration	• prolonged exposure

Arm & Shoulder	Possible CTD Problem
ARM WORK	
• exertions > 5 lbs	• little rest between exertions • large exertion
STATIC LOAD	
• prolonged holding	• unsupported • large exertion
ELBOW POSTURE	
• fully flexed	• repeatedly
• fully extended	
• rotated forearm	
SHOULDER POSTURE	
• flexed	• flex/abduct > 90°
• extended	• any extension
• abducted	
MECHANICAL STRESS	
• sharp edges	• repeatedly
• hard surfaces	• high force

Neck	Possible CTD Problem
NECK POSTURE	
• bend/twist > 20°	• > 50% of time

Back	Possible CTD Problem
LIFTING	• > 90% of AL • non-NIOSH
TORSO POSTURE	
• torso bending	• bend > 45°
• torso twisting	• bend > 20° + twist
STATIC HOLD/CARRY	
• > 5 seconds	• > 10 lbs • 5-10 lbs with flexed shoulder
STATIC LOAD	
• not able to change sit/stand posture over work day	• poor posture
PUSH/PULL	
• whole body action	• poor conditions

Legs	Possible CTD Problem
FOOT ACTUATION	
• foot pedals	• excessive force • extreme posture • high frequency or duration
LEG POSTURE	
• knee • ankle	• deep squat • kneeling • 1 legged posture • walk/stand on uneven surface
MECHANICAL STRESS	
• localized pressure	• high force
• kicking	• prolonged exposure

Figure 7-5. Factors and body parts to check in an ergonomics analysis.

- parts
- environment

In an ergonomics analysis, each involved body part must be checked for each of the factors listed in Figure 7-5; the figure also shows the point at which that factor could possibly cause a CTD problem. Figure 7-6 shows a simple form to use to track possible CTD problems.

Company. . . . .

Dept. . . . .

Supervisor . . . .

Job Name . . . .

. . . .

Date . . . .

Time . . . .

Pace:	mach.	self
Job Rotation:	yes	no
Regulator Operator	yes	no

Tools/Parts Weight

. . . .

. . . .

. . . .

Comments

. . . .

. . . .

. . . .

. . . .

. . . .

. . . .

. . . .

. . . .

. . . .

. . . .

. . . .

. . . .

. . . .

. . . .

CHECK IF A POSSIBLE CTD PROBLEM

LEFT HAND & WRIST	
Grasp	
Wrist Posture	
Frequency	
Mechanical Stress	
Vibration	

RIGHT HAND & WRIST	
Grasp	
Wrist Posture	
Frequency	
Mechanical Stress	
Vibration	

ARM & SHOULDERS	
Arm Work	
Static Load	
Elbow Posture	
Shoulder Posture	
Mechanical Stress	

NECK	
Neck Posture	

BACK	
Lifting	
Torso Posture	
Static Hold/Carry	
Static Load	
Push/Pull	

BACK	
Foot Actuation	
Leg Posture	
Mechanical Stress	

Figure 7-6. Form to track CTD problems.

INCIDENT RECALL TECHNIQUE

The incident recall technique (IRT) is little used in safety. But it has great potential. It is helpful to the line manager and the corporate safety program because it unearths many more accident causes than most techniques and finds these causes before they result in accidents of a serious nature. Furthermore, the IRT involves workers and demonstrates to them management's sincere interest in creating a safe working environment. For these reasons, seriously consider using this technique.

Confidential attitude surveys conducted by consulting firms in a number of companies have revealed that it is fairly common for supervisors to hide accidents. It is even more common for employees to hide them. The reasons usually given for not reporting accidents include the following:

- fear of discipline
- concern about the accident record
- concern about reputation
- fear of medical treatment
- dislike of medical personnel
- desire not to interrupt work
- avoidance of the red tape involved
- desire to keep personal record clear
- concern about what others might think
- poor understanding of the need to report

Most of these barriers to reporting accidents are eliminated by the IRT. It takes little effort on your part to gain considerably more incident information on causes through the IRT than through more conventional means.

The basic objective of the IRT is to gain the cooperation of the employee, so that he/she can and will freely relate all incidents from the past that he/she can recall. The success of the IRT, in terms of number of incidents revealed, depends primarily on the skill you use in the interview. The following steps have been effective in conducting the interview:

1. Put the employee at ease.
2. Explain the purpose of the interview and of the IRT.
3. Give assurance that the IRT is totally confidential. Your initial success depends to a great degree on this confidentiality.
4. Point out the benefits of the IRT to everyone—to the employee,

his/her family, the department, the company.

5. Show and explain the report to him/her.
6. Conduct the interview. Simply ask him/her to recall each near-miss accident that he/she has seen or heard about on the job. With each incident recalled, be sure to determined how many times he/she has seen it or heard of its happening. Jot down all pertinent information.
7. Ask lots of questions to fill in the gaps. Avoid interrupting him/her but get full information on each incident.
8. Review your understanding of the incident with him/her. Repeat it to make sure you've got it right.
9. Discuss the causes of the incident with him/her and possible remedies. Make it clear you want and need his/her help.
10. Thank him/her at the conclusion of the interview.

The benefits are obvious. First, and foremost, you may be able to prevent accidents from occurring. Second, you have heightened employee interest in safety by asking for and receiving employee involvement and participation. You have also gotten more information on current accident causes than you would have through a more conventional method. And you have demonstrated to the employee that you care, that you are sincerely interested in his/her safety. Furthermore, the IRT can help you check your safety sampling findings and give you additional input for your safety training programs.

The five techniques discussed can help improve the safety program and employee involvement in safety. In addition, organizations might use other common techniques, such as the following:

- the new type of safety committee that goes beyond merely reviewing injuries and complaining about physical hazards
- ad hoc committees formed for a short period to solve a specific problem
- task forces, often cross-sectional, to focus on a particular area of concern
- involvement in day-to-day organizational operations, such as inspecting, training, observing, etc.
- the one minute manager philosophy of involving the workers in the objective-setting process
- true situational leadership, where the amount of involvement varies with the workers' need to be involved
- behavior-based safety

REFERENCES

Petersen D. *The OSHA Compliance Manual.* New York: McGraw Hill, 1975.

Petersen D. *Safety Supervision.* DesPlaines, IL: American Society of Safety Engineers, 1999.

III

Options with Change of Structure

8 Self-Directed Work Teams

You can clearly get involvement, even beautiful involvement,without changing the organizational structure by just asking people to help make things better and by using such techniques as quality circles, safety improvement teams, ad hoc committees, or just simply asking for help. You can also achieve involvement and participation by changing the organizational structure. This may, however, be more difficult and, to a degree, more dangerous. In self-directed work teams, we are shifting the power in decision making and problem solving away from the boss and to, not the worker (as in empowerment), but the team of workers.

Description

A self-directed work team is a group that contains all the essential capabilities and responsibilities necessary to do a whole piece of work. Self-directed teams typically make decisions about methods of work, task schedules, and assignments.

For this type of system to work, the following must apply:

- Members must understand and buy into the goals of the team.
- The team must use the different capabilities of the members.
- There must be flexibility and sensitivity to the needs of others.
- The team must examine its processes continuously to learn and improve itself.

- The climate must be encouraging and nonthreatening with trusting relationships.

Self-directed work teams that are successful have the following characteristics:

- Team leadership is decided by the team.
- The appropriate person leads, based on skill and needs.
- The team assigns work to members by development need and skill.
- The team members are supportive of each other's developmental needs.
- Peer feedback is shared routinely, based on results and behavior.
- Team members train each other, and are continually improving existing capabilities.
- Capability development and ongoing competence is measured by performance.
- The team as a unit and individual members are committed and aware of performance.
- Each member makes a valid contribution.
- Team work is seen as a tool for accomplishing tasks.
- Formal and informal communications support a results focus.

Obviously the concept of teams severely alters the role of management, particularly the role of supervisors. The major shift in focus is from directing, assigning, controlling, problem solving, etc., the core work to resourcing the teams. Since the teams have the competencies needed to carry out most of their functions, supervisory and managerial roles are restructured to provide buffering or mediating changes in the environment so that the units are not upset as they carry on their work.

The new managerial functions might include the following:

- building appropriate capabilities within the units
- allocating needed resources to be used by units in support of achieving their goals
- auditing team performance to provide feedback and support
- evaluating the accomplishments of whole units, not individual members
- developing participation in setting goals and standards
- developing problem-solving capabilities in the units

The traditional role of a supervisor is to supervise one-on-one. In the team approach, the supervisor can end up in several rather different roles. He or she could be a team leader. While maintaining contact with each individual, the group leader may also share control and problem-solving duties with the group.

The supervisor can become a team coordinator. Various team members may take on specific leadership roles—initiating, seeking information, or consensus testing. The team coordinator manages the group primarily through coordination of their skills and activities. He or she also maintains contact with other groups and becomes involved in activities and projects outside the immediate group.

He or she could become a team manager, removed from the group and no longer with one-on-one contact with individual members. The team manager manages the group as a unit while maintaining full responsibility for the group's activities and production. The manager collects data and gives feedback to the group. As the individual held accountable for the group's output, he or she reenters the group when production does not meet expectations in order to diagnose problems and make necessary modifications in work patterns.

The supervisor could become a team resource person, available to a work team for any consultation or help that the team identifies itself as needing. The supervisor could move into any of these roles, or evolve from one to another (see Figure 8-1).

As a functioning organizational unit, the team has accountability for its own work. In essence, the first line supervisory role no longer exists. Various leadership patterns may emerge within the work team, but they are the responsibility of the team, and not dependent on external power.

In *Work Redesign in Whole Systems Design,* Richard Hackman and Greg Oldham describe the supervisor's new role in team environments:

> *Research has demonstrated that installing teams can cause serious strains in the relationship that exists between them and their supervisors, and numerous case reports have shown that such relationship problems can lead to an early and unanticipated demise of even well-conceived change projects.*
>
> *Reducing a supervisor's autonomy and power often prompts overcontrolling, rules-minded, and excessively critical behaviors on the part of the supervisor. Such behaviors can more than offset any positive changes in employee motivation and satisfaction.*
>
> *How might such counterproductive behaviors by supervisors be avoided when work is redesigned? The usual approach, of course, is to send the affected supervisors off to a training program where they would learn how to behave in ways that are constructive and supportive of their subordinates.*

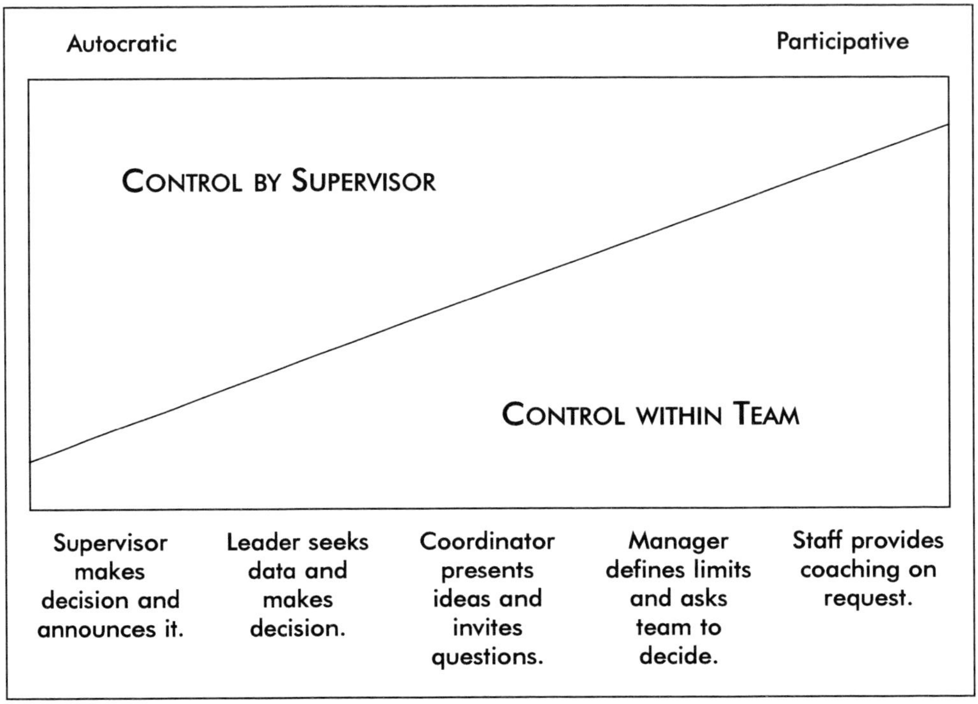

Figure 8-1. Supervisor roles.

The problem is that such programs have been shown to be largely ineffective in generating lasting changes in managerial attitudes and behaviors.

What may be required, then, is to change the duties of the supervisors, so that their own work grows in meaning and responsibility, and so that supporting autonomous work by their subordinates becomes a natural part of their job. Among the tasks that could be folded into supervisory jobs after work redesign are the following:

- *Gathering data for strategic analysis.*
- *Training employees in the new responsibilities and counseling them about both work-related problems and career opportunities.*
- *Helping teams and team members set performance goals and reviewing with them their performance in attaining those goals.*
- *Providing increased openness of communication both upwards (sharing employee concerns with higher management) and downwards (sharing with employees information about changes in organizational objectives and policies).*

- *Developing and testing with subordinates innovations in methods and procedures for executing and coordinating the work.*
- *Working on aspects of the work context (such as compensation and control systems, organization structures, equipment, and space) that may be causing dissatisfaction or impeding employees' work.*
- *Managing the evolution of the team development process itself.*

If supervisors are to perform these tasks well, their own jobs must contain considerable responsibility, discretion, and control. Since these qualities would have to be pushed down to the supervisors from higher levels of organizational management, the net effect would be to enrich the supervisors' jobs—just as previously had been done for the team members.

Such changes in supervisory jobs should prompt more supportive and effective behaviors by supervisors for at least three reasons. First, supervisors now would have more power to help their subordinates in meaningful ways.

Second, because of their new responsibilities, the supervisors would have less free time to closely supervise. In fact, there might be so many demands for the time of supervisors that they would be forced to give even more freedom and discretion to their employees.

Finally, because the supervisors would be integral members of the management team, they should have a greater investment in the success of the redesign effort. This should be the result if supervisors have been trained in the skill of coaching (rather than bossing) their subordinates and if they have been personally involved in the design of the teams.

(Reprinted with permission from R. Hackman and G. Oldham, *Work Redesign in Whole Systems Design*, W.W. Graham & Co., Inc., 1996, pp. 91-93.)

EFFECTIVENESS

The power of self-directed teams lies in the intrinsic motivation of the worker. Intrinsic motivation can be influenced by the design of work the individual is required to do, or what is commonly called their job.

Hackman and Oldham have developed a model that illustrates how certain job factors impact motivation. First, they identified the following three critical psychological states that impact work motivation (Figure 8-2):

- Meaningfulness of the Work implies that the performer of the work believes what he or she does "counts" for something in his or her value system.
- Responsibility for Outcomes occurs when an individual believes that his/her

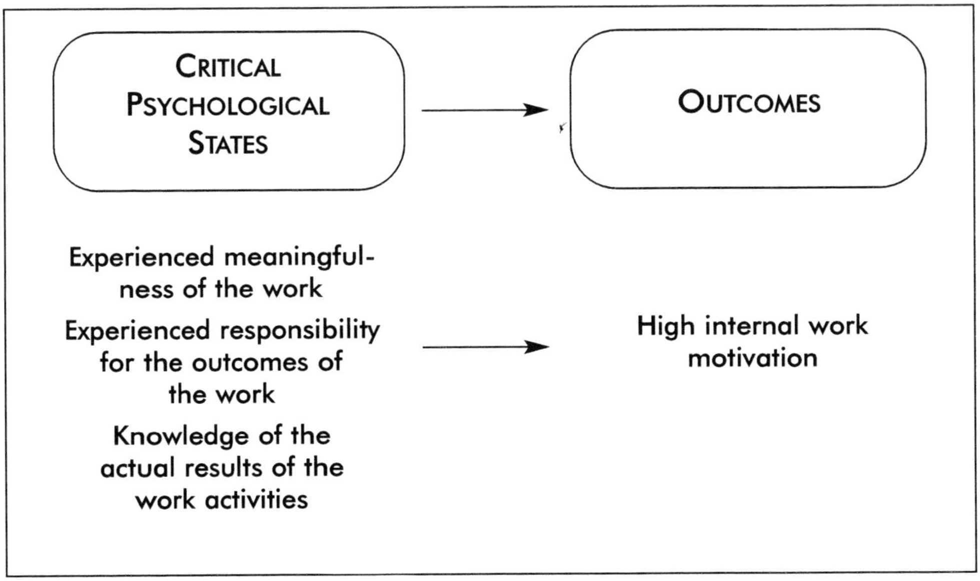

Figure 8-2. The three psychological states that affect internal work motivation.

(Reprinted with permission from R. Hackman, G. Oldham, Work Redesign in Whole Systems Design, *Tempe, AZ: W.W. Graham, 1996, pp. 91-93.)*

efforts or initiatives determine the quality or volume of the output.

- Knowledge of Results allows the job performer to assess his/her own performance and determine his/her satisfaction level, which results from his/her efforts.

They identified five primary job characteristics that directly foster these three psychological states: skill variety, task identity, task significance, autonomy, and task feedback. These five job characteristics and their impact are depicted in the enhanced model in Figure 8-3. The first three factors interact to create work meaningfulness, while the remaining two factors each related to the other two states.

- Skill Variety: The variety of different activities, and the requisite skills and talents to accomplish them.
- Task Identity: The "wholeness" of the job, the degree that there is a complete piece of work, with a clear beginning and end, leading to a visible product.
- Task Significance: The belief of the performers regarding the impact their own work and the result it has on other people.

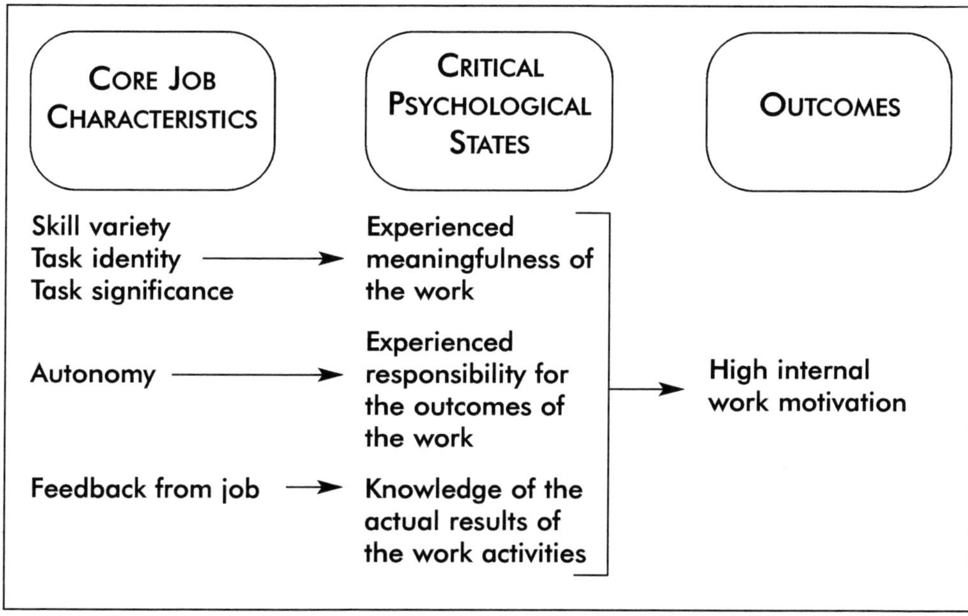

Figure 8-3. Job characteristics that foster the three psychological states.

(Reprinted with permission from R. Hackman, G. Oldham, Work Redesign in Whole Systems Design, *Tempe, AZ: W.W. Graham, 1996, pp. 91-93.)*

- Autonomy: The relative freedom to exercise discretion in how and when to do the work.
- Task Feedback: The feedback that the work activities themselves provide to the job performer.

Hackman and Oldham have concluded that these five job characteristics will determine the motivating potential of a job design, but that this potential will be influenced by several conditioning factors, including context satisfaction factors (pay, supervision, working conditions), individual desire to grow on the job, and capability of the individual to do the work.

Because team concepts provide much more of these five job characteristics to employees, self-directed work teams can be highly acceptable and motivational to employees.

Reference

Hackman JR, Oldham GR. *Work Redesign in Whole Systems Design.* Tempe, AZ: W.W. Graham, 1996, pp. 91-93.

9 Self-Directed Work Teams in Safety

In the last chapter, the concept of self-directed work teams was discussed in general. In this chapter we will specifically look at this concept as it relates to safety performance.

In a traditional but worker-empowered organization, little is changed with regard to safety. We still have clear accountability within the management structure for proactive daily performance of those things that prevent accidents (although workers may perform many of the duties). But accountability remains clear—it remains in line management from CEO to first-line supervisor.

In the team concept, however, things are changed. First. there is no line supervisor to be held accountable through the traditional management tool of measurement coupled with rewards. Under the self-directed work team structure, when it comes to safety, there *are* other things to be considered.

Traditional Safety Principles in High-Performance Settings

A few traditional safety principles need to be revisited in relation to a self-directed work team environment:

- The foreman is the key man (Heinrich, 1931): Today, there is no foreman.
- The three Es of safety (adapted from Heinrich 1931)—engineering, education, and enforcement: Two of these were traditionally the foreman's responsibility.

- Safety should be managed (Petersen, 1970).
- The key is accountability (Petersen, 1970): How do we measure and reward a team?
- Management starts with a clear definition of roles: What is the role of the self-directed work team?
- Supervisory tasks: Who will do the traditional supervisory tasks of inspecting, investigating, motivating, training, one-on-ones, behavior observations, etc.?

ATTEMPTS TO DATE

As we have jumped into high-performance approaches, we seem to have opted for one of several choices, based on the following assumptions:

1. We assumed that the teams would automatically take care of safety along with everything else.
2. In many cases, we also assumed that by merely removing a person (the supervisor) teams were ready for self- (team) management.
3. We assumed that we in management should treat the team as we would a supervisor, saying that "You are responsible for your own safety; now do something about it."
4. We assumed with employees now in charge of safety, that management was "off the hook," and no longer had a role -they abdicated.
5. We assumed that all people were ready to "take over." Some were not. Some people are happy just to do their thing.
6. We assumed that all employees want a piece of the action. Some did not. We established a three-tiered organization: (1) management, (2) involved workers, and (3) uninvolved workers, creating another friction point in some places.
7. We sometimes assumed that if it did not work, we could always go back and take back the authority; but when we did that, it resulted in considerable chaos (a mild description!)

HIGH-PERFORMANCE SUCCESSES

Occasionally we found that the approach worked brilliantly, resulting in unheard of success. Having noted some of the pitfalls, now we will discuss how to achieve success with team approaches.

Many organizations have been highly successful in utilizing the team approach. I have seen small units that have made the transformation with remarkable results. In talking with team members in successful companies in a variety of industries (food processing, oil drilling railroads, steel companies), I have never seen such excitement from hourly workers as with those experiencing team participation. And, without fail, the resultant safety records have been superb.

I also have seen the opposite—organizations where going to team approaches has produced problems and discontent, reduced morale, increased friction and occasionally created utter chaos.

What are the differences? Probably a number of things: readiness for participation, the planning that went into the change, the amount of involvement the workers had in the change decision and process, and so forth.

In those organizations that are successful, there are some things that management seems to do. Listed here are suggestions that any organization might well consider following:

1. Define within management exactly where you are willing to go by defining what you mean by participation. Is it to get better input so management can make better decisions? Is it to share decision making? Is it to turn decision making over to teams? And which decisions will management retain?
2. As best you can, check the maturity level of both management and the workers. Find out if you are "ready." This can be done through interviews or perception surveys. Also, look at the degree of self-management you have allowed in the past, before going to teams.
3. Check the amount of confidence and trust that exists between management and the workers. Culture surveys can help in the assessment.
4. Clearly define the ground rules, the guidelines defining what is allowable in decision making and what is not (regulatory compliance, core corporate values, etc.).
5. Allow great flexibility within those guidelines. For instance, "You must have a system in your team of behavior observations. How you do it is up to you."
6. Require each location/team to have an annual (or periodic) plan in safety. Have an annual (or periodic) check of each unit to see whether or not it is carrying out the plan.

7. Hold each location/team accountable for what it has agreed to do.
8. Determine the core elements that should be dealt with in each plan, but allow great flexibility in what each location/team can do to achieve results within those core elements.
9. Require intermediate self-checks at shorter intervals, making frequent adjustments as needed.

Barbara Trautlein and David Milner described the changes in safety when an organization shifts to a high performance mode:

> *New systems of work organization, such as team-based environments and innovative employee compensation and reward schemes, signify a paradigm shift in how workplaces approach safety. Within these systems, safety is no longer viewed as an add-on program or afterthought, second to quality and production demands; rather, safety is designed into the system, becoming an integral part of how the system is operated. Teams of self-directed workers accept increased responsibility for personal and workplace safety. Safety concerns are addressed before problems occur, which allows organizational resources to be redirected from reactive activities such as regulatory compliance, accident investigations and record keeping. Such an approach transcend basic legal requirements and stretches into the realm of paradigm shift. The bottom line: Safety truly becomes a way of life, embedded into organizational structure.*
>
> ### *Advantages of the New Paradigm*
>
> *The role of safety in a self-directed work environment differs greatly from its traditional role. Although such an environment presents some challenges, its benefits greatly outweigh its risks. If a safety professional manages these opportunities effectively, the entire organization benefits.*
>
> *In a traditional organization, safety is reactive and viewed largely as the safety professional's responsibility. Typically, first line supervisors are held accountable for their employees' safety performance. This approach has met with limited success.*
>
> *In contrast, the continuously improving organization and safety program utilizes self-directed teams to perform "traditional management" duties. The strength of self-directed teams is that they provide decision-making authority at the most appropriate level. This speeds the improvement process, fosters employee acceptance and produces an improved end product.*

Before assuming this authority, teams must be prepared. Here, the safety professional plays a key role. He/she must transfer knowledge to the work teams, thus creating many safety "agents" on the "firing line" where accidents and unsafe conditions occur. This requires close work with the teams, as their empowerment leads to increased responsibilities. Eventually the practitioner is viewed as a resource, someone who provides compliance training, timely safety information and the analysis.

Incorporating safety into work procedures is a key element of the self-directed team approach. Safety actions are listed with quality and operational instructions, a step that elevates safety procedures to equal performance status. These responsibilities (which include OSHA mandated training) are then incorporated into a skill-based pay (SBP) system, which helps motivate team members to actively participate and demonstrate their safety training. Some develop sufficient knowledge and skill in a certain subject areas, allowing them to become staff "training experts," which frees the safety professional to address other issues.

In addition, teams self-evaluate their adherence to work procedures and initiate immediate corrective actions when needed. This process is more effective than waiting for a safety audit.

Preliminary accident investigations are another work team function. In many cases, teams skillfully identify root accident causes and eliminated them. In a self-directed environment, the safety professional serves as a consultant to the team during investigations, but does not usurp its responsibility.

Teams also initiate and track accident efforts. The safety professional advances these efforts by soliciting management cooperation and obtaining necessary funds. Similarly, teams inspect work areas using inspection checklists developed by the safety department. The safety professional's role in these efforts involves providing training and implementing corrective actions.

In a self-directed work team organization, the safety professional focuses on training, education and communication. Traditional "policing" is not required; teams discipline themselves. Teams realize that safe work procedures (which they help develop) must be followed, and that they are responsible for personal and workplace safety. To facilitate team efforts, the safety professional conducts compliance audits and training programs, and provides information about new legislation. The end result: by transferring knowledge to team members, the safety professional can focus on conducting research and improving processes, rather than on administrative duties associate with compliance.

Creating a new system of work organization—whether a greenfield, start-up facility or plant redesign—produces a multitude of possibilities for re-inventing each aspect of an organization, particularly how the safety function is perceived, managed and operated. Whether designing a total manufacturing and staff support system under sociotechnical principles, focusing on the factory floor, or creating a skill-based pay (SBP) system, opportunities for altering an organization's approach to safety abound.

The first design principle is to "design for purpose." What goals does the company wish to achieve? Focus on the company's overall mission and values: What business and people outcomes are paramount? Mission and values statements convey what is unique about the company. The entire design, as well as each design recommendation, must be based on and support these goals. Safety must be explicitly stated. From these company-wide goals, departmental, team and individual goals and responsibilities are derived.

The second program design principle is "design by end users." Typically, design projects consist of a steering committee and a design team. Comprised of upper level management (and labor officials if unionized), the steering committee sets the boundaries for the design, provides resources and grants final approval of recommendations. The design team is a working body. After receiving its charter from the steering committee, the team creates proposals. This process involves seeking input from appropriate stakeholders, gathering organizational data and benchmarking successful systems. to design for safety, the team must include employee representatives, maintenance personnel, first line supervisors, staff support engineers and safety personnel.

Teams of self-directed workers assume increased responsibility for safety in the new work system. Their decision-making authority, which involves such traditional management functions as daily task assignments and personnel scheduling, explicitly includes safety responsibilities. Team communication mechanisms place safety issues high on the agenda. For example, during daily preshift meetings, when 24 hour quality and production performance records are received and task assignments made, safety concerns are discussed and follow-up activities assigned. During loner, periodic (weekly or monthly) team meetings, long-term, in-depth safety issues are discussed.

Teams often elect a safety leader who is accountable for accident investigation, reporting and feedback. The leader may also serve as the team's safety trainer. Utilizing a "train-the-trainer" approach, the safety professional teaches this individual presentation skills and provided all training materials. The leader then trains team members and maintains required documentation. because the leader is also a team member, s/he provides constant reinforcement and feedback on a real time basis—when it is both most needed and most effective.

Safety goals designed by self-directed teams must be in concert with departmental goals and company mission and values. Additionally, teams receive (or design) tools to track and monitor their performance relative to these goals. Based on both feedback about this performance and the skills required to perform their roles, teams conduct annual training needs assessments. They determine what training/certification/recertification (i.e. hazardous materials handing, first responder, etc.,) courses members need during the next yeas. The assessment then becomes part of the company's annual business plan.

Finally, teams are self-disciplined; they must ensure that each team member follows safe work practices. Teams employ various tools to facilitate internal compliance with safety rules. (When such methods fail, however, management is the last line of defense. Legally, employee safety is ultimately management's responsibility.)

As teams assume greater responsibility for safety, individual accountability also increases. Floor employees (both operations and maintenance) receive new role descriptions, which transcend traditional job descriptions. They include responsibilities for overall product and technical quality, as well as social and business results. These descriptions clearly convey that team members, not only line supervisors, are responsible for "running the business." Safety responsibilities are integral parts of these descriptions.

Although further removed from daily operations and focusing on external linking and long-term goals, line management shares accountability for safety. Labels such as 'supervisor" and "foremen" are replaced with "resource" or "facilitator," which reflect the new role as coach (rather than boss). Line management fosters a safe working environment by supplying resources to resolve safety problems, helping teams deal with rule violators and providing leadership (as opposed to micro-management).

Less successful work systems fail to consider staff functions as relevant stakeholders; thus, these individuals are never asked for input during safety organization design. Safety personnel and staff engineers are left to work and be evaluated in a traditional manner, yet are expected to interface with and provide service to team members and line personnel in nontraditional

> *ways. Therefore, safety personnel's roles must be broadened beyond accident investigation and compliance activities. Similarly, when selecting/designing/redesigning equipment and processes, engineers must seek input from line personnel and end users.*
>
> Source: B. Trautlein, D. Milner, "Work System Design, Creating a New Paradigm," *Professional Safety*, December, 1994. Reprinted with permission.

The largest problem that organizations have when shifting to team-run safety is the issue of accountability. How do we hold a team accountable? Neill Carson writes to this point:

> *Ask any manager about accountability, and s/he'll tell you it's a great thing to have. Ask him/her exactly how accountability works, however, and the answers start to get vague. The picture gets even hazier as more organizations move to self-managed teams, autonomous work groups and cross-functional teams.*
>
> *In the rush to adopt these new models, the concepts of accountability sometimes gets relegated to the dustbin of "old ways of thinking"—one more relic of the 'antiquated hierarchical system." Now the central question refuses to go away: When things go wrong—or, just as important, when things go wonderfully right—who gets the blame, the credit or the consequences?*
>
> *Hold on, you say: In a team based organization, there aren't supposed to be any accountability problems. Once people are assigned to work teams, everybody naturally becomes committed, motivated, interested and eager to contribute to the company's success. Isn't that better than the old system of bureaucracy, blame, buck passing and CYA?*
>
> *As some companies have demonstrated, it may indeed be better. But as others have discovered, it also may be worse. Only one thing is certain: You can obscure the issues around accountability, but you cannot avoid them. You either have accountability, or you don't. Most companies think they do...but they don't.*
>
> *...We have a pretty good idea of what it means to hold an individual accountable, but does the work have the same meaning when we apply it to a team? Have you ever heard of a team being fired for doing a poor job? Ever heard of a team being promoted for doing an outstanding job? Certainly, bonuses are handed out (or not), but everybody gets (or loses) the same amount. Does that reflect the differences in contributions from the team's members? Most of the time it does not. And what are the implications for*

the best workers? When there's little recognition for outstanding effort or ability, would you expect them to put forth their finest efforts the next time around? The grapevine doesn't provide a basis for compensation decisions.

None of this means we should scrap the teams we've got or stop forming new ones. There is plentiful evidence that work teams, project task forces and other collective entities can accomplish far more than individuals working separately. But much of the potential of teams is wasted unless it is harnessed through an appropriate structure and monitored through individual accountability.

...The conditions for managerial accountability apply equally to the leaders of ad hoc teams, task forces and so on. For a team leader to be held accountable for the work of the team, s/he must have:

1. *The right to veto any member of the team.*
2. *The right to decide on work assignment in the team, and to hand out the tasks 3. The right to send recommendations and evaluations back to the team member's "real" manager. (The team leader can recommend bonuses or merit increases, but usually can't decide on them. That's the manager's job.)*
3. *The right to initiate (with due process) the removal of any member of the team.*

We are verging on an accountability crisis today that goes beyond work teams. Whether flat or layered, Fortune 500 giants or entrepreneurial start-ups, companies often lack a clear and understandable structure that makes managers and employees accountable for their work. The fear that individual leadership and accountability will somehow suppress or destroy team effort and participation is groundless. In fact, the opposite is true. Ambiguous accountability damages morale and leads people to seek political solutions to problems instead of business solutions. Accountability is a fundamental principle of business success and must be designed into the management structure. When accountability is clear, productivity and efficiency follow. That's the aim of all good managers and good companies.

The Proctor and Gamble Company has utilized team approaches and high employee involvement for many years, with excellent results. While turning over much of the safety effort to employee teams, they have set the ground rules for locations/teams/units by identifying key elements that teams must deal with. Eugene Earnest, formerly with Proctor & Gamble, describes in the following excerpt one of the more successful team safety approaches.

About twelve years ago, the Proctor and Gamble corporate safety group developed what is known as the Key Elements of Industrial Hygiene and Safety [IH&S]. In effect, these are the what counts activities for IH&S and are the basis for site surveys. It was believed that if these activities were effectively implemented, injures and illness would be reduced and conversely, if they were done poorly, injuries and illness would increase. Because line management was involved in the development of this list, there was "buy-in."

I cannot stress enough the importance of having a clearly identified IH&S program against which goals can be established at all levels of the organization and people held accountable for before the fact measures of injury and illness prevention. I recognize that certain areas of safety are "soft" and to not readily lend themselves to measurement; however, injury- and illness-producing events will invariably be linked to behavior (of management and/or nonmanagement) which is observable and measurable. That which is being measured will vary depending upon the role of the individual. Following are examples of behaviors that could be measured at plant manager and department manager levels.

Plant Manager

- *Expectations*
  Sets clear expectations of his/her group managers in safety management
  Wants to review goals and forecasts
  Wants to hear about safety on every hour
  Wants to hear or read about safety program during daily, weekly, monthly reports from the group manager
  Wants to see proactive behavior of his/her group manager in safety

- *Challenges*
  Goals-from modules, departments
  Forecasts-from module mangers
  Results from module managers
  Unsafe conditions, equipment design, safe practices

- *Behavior Feedback*
  Confronts any unsafe behavior no matter where he/she is going or what he/she was planing to do
  Confronts management behavior that needs to be recognized or corrected to improve the safety performance

- *Presentation*
  Reviews expectations with group managers and department managers
  Discusses safety in presentation to managers and sometimes technicians
  Incidents
    -involved in all recordable incident reviews
    -involved in all serious property damage reviews

- *Resources*
  Works staffing plans to ensure that effective and respected managers are placed in the safety position
  Supports operating and capital expenditures to achieve results

Department Managers

- *Expectations*
  Informs his/her people and peers that
    -they do not have the right to work unsafely
    -they will be confronted for unsafe behavior, recognized for safe behavior, and expected to confront other's safe behavior
    -they are expected to contribute to the safety program by setting strong personal example, identifying and correcting problems

- *Goals*
  Develops department goals and action plans to met those goals

- *Forecasts*
  Prepares monthly forecasts of results in the department
  Critiques pervious months' performance vs. forecast

- *Behavior Feedback*
  Confronts any unsafe behavior no matter what he/she is doing or what he/she was planning to do
  Recognizes safe behavior

INDUSTRIAL HYGIENE AND SAFETY ELEMENTS		
Element	Profile Rating	Priority
I. Organizational Planning and Support		
A. Expectations and Involvement		
B. Goal Setting/Action Planning		
II. Standards and Practices		
A. Standard Implementation		
B. Safe Practices		
C. Planning for Safe Conditions		
III. Training		
A. Site Training Systems		
IV. Accountability and Performance Feedback		
A. BOS/Safety Sampling		
B. Behavioral Feedback		
C. Performance Tracking		

Figure 9-1. Key elements.

- *Presentation*
  Discusses safety topics in presentations to technicians and managers

The key elements of IH&S referred to previously are divided into four major categories and nine subcategories as shown in Figure 9-1.

Each key element is rated by the surveyor utilizing a scale of zero to 10 where zero means "Nothing has been done" and 10 means the key element is "Fully implemented and effective." Calibration to ensure that assigned ratings on surveys do not vary plus or minus one point is accomplished by corporate liaison accompanying the business sector surveyors on select surveys worldwide. Training and qualification programs for site health and safety resources also include exercises to ensure that people entering the role fully understand the Industrial Hygiene and Safety key elements and are calibrated in assigning ratings.

The validity of this approach is evident from the graph in Figure 9-2 that illustrates the correlation between key element ratings and total incidence rates.

New acquisitions have typically had total incidence rates (TIRs) of 12

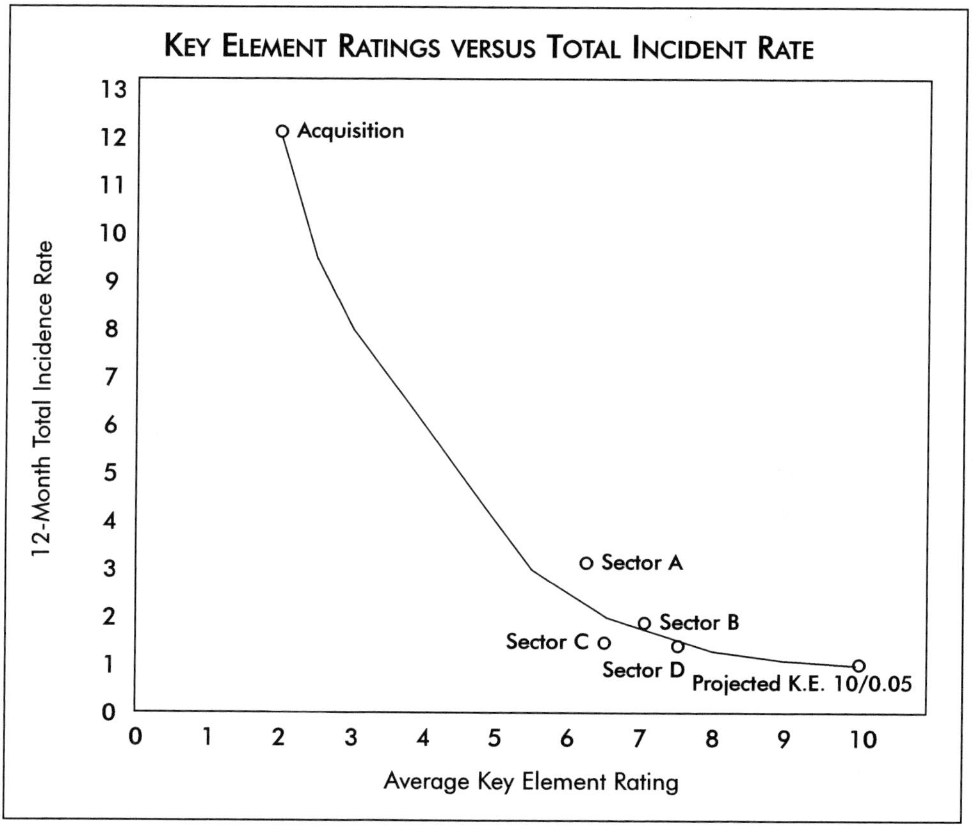

Figure 9-2. Key elements versus total incident rate.

and above and Key Elements ratings on the initial survey of two to three. Established business sectors (A through D) have considerably higher Key Elements ratings and correspondingly lower TIRs. The process of phasing in acquisitions typically requires several years. The projected TIR for a facility having a key element rating of 10 is 0-0.5.

Although it is not within the scope if this article to provide a detailed explanation of each key element, it should be noted that Organizational Planning and Support, Training and Accountability and Performance Feedback above are primarily behavior-related as well as Safety Practices under Standards and Practices. Although OSHA standards are grouped and rated under Standards and Practices, other key elements also reflect how well they are being managed. The arrangement of the Key Elements provides a tool that heavily reflects behavior while covering the full program scope. The Priority column on the chart is used by the surveyor and the site to establish focus areas for improvement. Generally this list should be limited to three items so as not to overwhelm the site.

I now would like to draw your attention to one of the nine key elements and share some learnings. As noted previously, a behavior observation system (BOS) is essential to a behavior-based safety management program and is therefore the element I want to discuss. There is a saying, "Anything worth doing is worth doing well." This is particularly true when dealing with the design and implementation of a behavior observation system. A BOS that is introduced to a site without giving close attention to design as well as preparation of the organization to fully understand and own this powerful tool can lead to failure and require considerable effort by the organization to get it reestablished. Many of the learnings from BOS came from sites that have struggled with the key element and won in the "quest" for conscious improvement.

Following are characteristics I have noted at sites that have implemented effective behavior observation systems.

1. Observations focus on critical behaviors, not behavior outcomes.

It is not unusual to find a department BOS degenerate into a system that is skewed to behavior outcomes. I find that there are several reasons why this may occur. First, the system owner may not have a clear understanding regarding the importance of focusing on behavior rather than behavior outcome or may not have recognized the difference between the two. You cannot confront or reinforce a condition, only a person; therefore, immediate feedback opportunities are lost if the system focuses on outcomes. Furthermore, you may not know who caused the condition and therefore feedback can only be given to a group rather than an individual. Another reason I believe that behavior outcomes are frequently contained in BOS is the fact that it is less threatening to observe behavior outcomes (conditions) and therefore, the observer isn't obliged to provide verbal feedback.

2. The list of critical behaviors is kept manageable.

I normally recommend that the list be limited to five or ten critical behaviors. Generally 15 to 20 minutes will be required to make the observations. I have seen elaborate behavior observation systems that looked great on paper, but have failed because they became too burdensome. One such case involved a large manufacturing facility that had many departments each of which had about 50 critical behaviors identified. This system failed because feedback was delayed. In this case, all completed observation forms were sent by technician observers to a central location for computerized processing. The information was not processed in a timely manner causing the observers to question the importance the organization placed on the systems. Due to this unfortunate incident, considerable time and effort was required to reestablish an effective system at that location.

3. Behaviors are clear and specific so that anyone trained in BOS arrives at the same conclusion.

You would normally expect to find critical behaviors included in site safe practices. If complex behaviors are observed such as locking out certain items of powered equipment, this information must be sufficiently detailed to enable all observers to arrive at the same conclusions.

4. Behavior observations are conducted at least once per shift.

Infrequent sampling is not effective and utilizing the rating system referenced above would likely result in a rating of 4 by a surveyor "Only partially in effect-results unsatisfactory-much room for improvement." Sites with the most advanced BOS systems utilize control charts for each critical behavior as well as for overall percent safe behavior and number of observations per week.

5. An effective feedback system is in place.

This includes positive and corrective feedback to those being observed as well as to the work team. Characteristics of effective feedback include:

a. Immediacy: There should be minimal delay between performance and feedback concerning it.
b. Frequency: Generally, the more frequent the feedback the better
c. Specificity: Feedback should pinpoint the specific behavior(s) being addressed.
d. Understandability: The information being fed back to the performer(s) must be in a form that is easily interpreted.

6. There is broad ownership of the BOS.

Systems that are most effective have "buy-in" by both management and nonmanagement personnel and observations are conducted by both groups. Some of the better systems are conducted by nonmanagement personnel with full support from management by providing time and resources. On the other hand, systems totally owned by management have been less effective.

7. Injuries (long-term) correlate with behaviors.

If unsafe behaviors account to 88 percent or more of accidents, logic indicates that a correctly designed system that focuses on critical behaviors and is fully implemented should provide a reliable tool to predict injuries. If this is not the case, there is something wrong with the system and trouble shooting will be necessary.

... The Key Elements of IH&S have proven to be effective over time and as shown by the above correlation, and have provided tremendous leverage for sites worldwide to manage their Industrial Hygiene and Safety programs to a key element rating of 8+ (satisfactorily implemented and effective). If a plant has a key element rating of 8+ or better and the injury and illness total incidence rate is low it can be said with a high degree of certainty that they are truly doing a good job in managing safety.

(Reprinted with permission from R.E. Earnest, "What Counts in Safety?" *Insights into Management,* 2nd Quarter, 1994, pp. 2–6. Reprinted with permission of author and National Safety Management Society.)

The Proctor and Gamble system as described by Eugene Earnest shows that we can have accountability for safety performance in a team (employee-run) safety system. There is flexibility in the ways things may be done, but no question that they must be done.

REFERENCES

Carson N. "The Trouble with Teams." *Training.* August, 1992, pp. 38, 40.

Earnest RE. "What Counts in Safety?" *Insights into Management.* National Safety Management Society, 1994, pp. 2-6.

Heinrich H. *Industrial Accident Prevention.* New York: McGraw-Hill, 1959.

Petersen D. *Techniques of Safety Management.* New York: McGraw-Hill, 1970.

Trautlein B, Milner D. "Work System Design, Creating a New Safety Paradigm." *Professional Safety.* December, 1994.

IV

Culture and Involvement

10 Culture

Probably the best definition I have heard of culture came from a worker I interviewed. "Culture is what everybody knows and therefore does not have to be stated or written down."

The concept became a very popular management subject in the early 1980's, probably because of the popularity of the book *In Search of Excellence* by Peters and Waterman. That book described what it was that accounted for the economic success of a number of companies—what made them good. Other works followed, delving into the concept.

Basically the concept of culture had been considered long before that. In 1967, Dr. Rensis Likert wrote in *The Human Organization* about his research on the styles of different companies and how these styles affected the bottom line. Dr. Likert coined the term *organizational climate.* We now call that culture.

Likert not only researched organizational climate, he also evaluated it in ten areas:

1. Confidence and trust.
2. Interest in the subordinate's future.
3. Understanding of and the desire to help overcome problems.
4. Training and helping the subordinate to perform better.
5. Teaching subordinates how to solve problems rather than giving the answer.

6. Giving support by making available the required physical resources
7. Communicating information that the subordinate must know to do the job.
8. Seeking out and attempting to use ideas and opinions.
9. Approachability.
10. Crediting and recognizing accomplishments.

Dr. Likert invented a way to measure climate with a forced choice questionnaire, which he administered to employees of an organization to determine their perception of how good the company was in the ten areas. He later took the perception survey results and ran correlational studies with such measures as profitability, return on investment, growth, and other bottom-line figures, invariably coming up with extremely high positive correlations. Apparently climate determines results.

As culture (climate) became a popular management subject in the 1980s, executives began to look at their organizations and consider ways to "improve their culture," Many organizations displayed new posters on the walls describing "their culture." We know today that if the management of a company must write it down and make a poster of it, they are not describing their culture—they are describing what they would like it to be. No employees need the culture to be described to them—everybody knows "what it is like around here."

CULTURE AND SAFETY

The field of safety pretty much ignored the concept of culture throughout the 1980s. As management attempted to improve culture through changing styles of leadership, employee participation, and so forth, safety professionals tended to change their approaches very little, keeping the same tools, using the same elements in their safety programs that they had always used. Safety programs typically consisted of the usual things; meetings, inspections, accident investigations, using job safety analyses (JSAs), and so on. These tools were perceived as the essential elements of a safety program. OSHA published a guideline in the 1980s suggesting that all companies should follow all of these practices. A number of states enacted laws requiring companies to do these things. These traditional elements were regarded as a safety program.

While OSHA and the states were going down the essential element track to safety (as was much of the safety profession), a number of researchers began to suggest totally different answers to the safety problem. Most of their research results were consistent in saying that "there are no essential elements"—what works in one organization will not in another. Each organization must determine for itself what will work for it—there are no magic pills. The answer seems to be clear: it is the culture of the organization that determines what will work in that organization.

Certain cultures do, in fact, have safety as a key component, whereas other cultures make it very clear that safety is unimportant. In the latter almost nothing will work; meetings will be boring, JSAs perceived only as paperwork, and so on.

The culture of the organization sets the tone for everything in safety. In a positive safety culture, it says that everything you do about safety is important. In a participative culture, the organization is saying to the worker, "We want and need your help." Some cultures urge creativity and innovation, some destroy it. Some cultures tap the employees' for ideas and help; some force the employees never to use their brains at work.

Attributes of a Successful Culture

In *In Search of Excellence*, Thomas Peters and Robert H. Waterman suggested the following eight attributes that make up a successful culture:

1. *A bias for action.* "These best companies may be analytical in their approach to decision making," but they are not paralyzed by it. Standard operating procedure is "Do it, fix it, try it."
2. *Close to the customer.* "These companies learn from the people they serve... Many...got their best product ideas from customers. That comes from listening intently and regularly," (also listening to their employees).
3. *Autonomy and entrepreneurship.* These companies foster leaders and innovative-"champions"-by encouraging practical risk-taking and supporting good attempts; even when the attempts fail.
4. *Productivity through people.* "Excellent companies treat the rank and file as the root cause of quality and productivity gain," (and safety).
5. *Hands-on, value driven.* Excellent companies pay "explicit attention" to values and "their leaders have created exciting environments through personal attention, persistence and direct intervention-far down the line."
6. *Stick to the knitting.* "The odds for excellent performance seem strongly

to favor those companies that stay reasonably close to the businesses they know."

7. *Simple form, lean staff.* "The underlying structural forms and systems in the excellent companies are elegantly simple...Top-level staffs are lean...."
8. *Simultaneous loose-tight properties.* "The excellent companies are both centralized and decentralized ... they have pushed automation down to the shop floor or product-development team. On the other hand, they are generalists around the few core values they hold dear."

How Is Culture Built in an Organization?

The following are just a few considerations that determine the culture:

- How decisions are made: does the organization spend its available money on people? On safety? Or are these ignored for other things?
- How people are measured. Is safety measured as tightly as production? What is measured tightly is what is important to management.
- How people are rewarded. Is there a larger reward for productivity than for safety? This states management's real priorities.
- Is teamwork fostered? Or is it "them vs. us"? In safety is it "police vs. policed"?
- What is the history? What are the traditions?
- Who are the corporate heroes? And why?
- Is the safety system intended to save lives or to comply with regulations?
- Are supervisors required to do safety tasks daily? This says that safety is a big value.
- Do big bosses wandering around? Talk to people?
- Is using the brain allowed on the work floor?
- Has the company downsized?
- Is the company profitable? Too much? Too little?

ACCOUNTABILITY, CULTURE, AND BEHAVIOR

For a number of years we have been observing the need for accountability systems in management as an essential to safety success. This probably started in the mid-1960s or even before. Prior to that, we had pushed for management to be responsible for safety, probably stemming back to the 1930s from one of Heinrich's axioms.

While we have recognized that accountability is the key, in most organizations we never quite got meaningful accountability systems for safety performance in place, not being able to either accept or sell the fact that to achieve accountability requires new measures of performance and new reward structures.

Then came the culture era and the behavior-based safety era. Some organizations went into culture building (or talked about it), while others were into some variety of behavior modeling (or talked about it).

In both of these directions, there has seemed to be an unwritten law that either of these was the answer—nothing else was needed. The Culture Companies seemed to jump to a belief that once the new culture is in place, the accident problem would be solved; nothing else would be necessary. The new culture posters and pamphlets described the new utopia that now existed. We no doubt needed posters and pamphlets to convince the employees that they were now living in an environment where safety was key; was a core value. Often management believed things had changes; believed their own writings. Often the employees somehow missed the change—things were the same to them. The Culture Companies seemed to feel that since there was a positive safety culture, the old stuff was no longer necessary. Forcing managerial performance through accountability didn't fit anyone—too out of step with the new culture.

The Behavior-Based Companies knew that now that the employees were defining what safe behavior was, and were observing each other, obviously safety was solved, allowing their managers and supervisors or team leaders to concentrate on the truly important things like getting the product out the door, improving quality, doing paperwork, etc. These were the things that they knew to be important because these were what they were being held accountable for.

Not all companies went these two routes. Many kept down their traditional route: keep the safety program as they had always done it—an easier route. Others opted for traditional programs as outlined by government—an easier and safer route. Some bought packaged programs. And many did nothing.

In the above six routes, all could manage without management accountability. All say they have management's responsibility, but happily, managers and supervisors or team leaders do not have to do anything about safety.

To compound the situation we have fewer managers and supervisors today; longer span of control; fewer aides to help; more paperwork to fill out; and, in many cases, supervisors must spend much more time in the office on that great time-saver, the computer.

THE REAL FORMULA

What is the answer to safety system excellence? The behavior-based people are right—you cannot achieve safety excellence with out proper behaviors. But, the culture builders are right also, for you cannot get the proper behaviors without the right culture. And here's the real hooker—you cannot get the right culture without accountability. It is not either one or the other—you must build all three—and in the proper order to achieve excellence in safety.

Management accountability—a system of role definition, correct measures of performance and adequate rewards contingent upon that performance forces managerial and supervisory daily proactive activities to take place. These actions build a culture which states: safety is so important around here that all the managers and supervisors have to do something about it every damn day." When employees believe that the behaviors will follow—both behaviors like working safely and behaviors like becoming involved in helping the whole process.

Safety excellence can be achieved as long as we re-learn our alphabet:

$$A \rightarrow C \rightarrow B \rightarrow E$$

Accountability builds Culture which gets Behaviors resulting in Excellence.

ASSESSING CULTURE

Although you already have a sense of the culture if you have been with an organization a while, to get a truly accurate picture you must question the working employees—the hourly people.

In the book *Managing Employee Stress*, we suggested ways to assess organizational culture as a primary cause of job stress. The first technique is to get information on employee health disturbances (emotional, behavioral and health deterioration) and determine the trends of the employee population. One way of doing this is through the use of questionnaires asking anonymously about:

- tension on the job
- anxiety
- sleeping problems
- depression
- fatigue
- inability to concentrate
- aches and pains
- headaches

- satisfaction with job
- alcohol consumption
- increased smoking
- drug abuse

Information from other sources will tell you about the trends in:

- accidents
- absenteeism
- turnover

Information may (or may not) be available on trends in stress-related diseases of the work force.

Questionnaires might also be used to ask anonymously about such items as:

- job control
- importance of work
- relationships with the boss
- pressures from workload, time

Another information source is to look at such things as:

- repetitiveness of work
- machine speeds
- forced pace
- forced overtime
- downsizing
- isolation
- need for vigilance
- demands for preciseness
- people being rushed
- seriousness of mistakes
- unsafe behavior

You might look at what support systems are available to the workers within the company or in the area. In-depth interviews with hourly employees will often unearth much of what you need to know to assess the real situation.

Traditional safety measures do not tell us much. Numbers of accidents may or may not be an indicator. Accidents correlate very highly with stress-producing environments, but it takes an extremely large unit to generate enough accidents to attain statistical validity. In most cases they become relatively meaningless figures. Audits give little indication as they typically tell us little about the existing climate.

Interviews and surveys, however, attempt to find out from all levels of the organization what climate really does exist. If properly done, they are valid and very meaningful.

Asking employees can be very useful in assessing the real corporate climate, assuming they will level with you. Whether or not this happens is probably dependent on trust and spending enough time in the process to get to where they will level with you.

Often it takes an outsider to get to the truth from the employees. This fact in and of itself often suggests a problem exists. It also usually takes time to get to reality in interviews. I have found that in most organizations it takes 10-15 minutes of general talk during an hourly interview before anything important begins to emerge. It takes that long for trust to be established. Once established it often takes an hour or more to stop the out-pouring of facts and feelings that emerge. When this relationship is established, the interview is probably the best technique to get to reality. But do not expect good data from short interviews.

The second option is the survey. With this approach a lot of data can be gathered quickly and it can be processed rapidly by computer. But the survey instrument is crucial. It must be constructed by people who know what they are doing and that takes time. Each question must be carefully constructed and validated by a professional test development person working with subject experts. We have only recently been blessed with some general safety climate surveys (and only a few are currently on the market). (See Chapter 11, Assessing Culture.)

IMPORTANCE OF CULTURE TO INVOLVEMENT

Involvement will only happen in the right environment (culture). Before attempting involvement an organization needs to now exactly whether or not there is readiness (maturity) in both the hourly work force and in management. When an organization is not ready, attempting involvement will be a major mistake.

In recent years I have worked with a number of companies where the attempts to push ahead has led to pure chaos. In interviewing people in these organizations, I've collected some interesting quotes:

"After 100 years of a parent-child relationship, we're trying to get to an adult-adult." (Executive)

"Our success depends on the maturity level of our people—manager and worker." (Executive)

"First line supervisors are frustrated, lost and have a full plate now." (Mid Manager

"Roles are defined and yet supervisors say they don't know what to do about safety." (HR)

"Empowerment is a farce." (Hourly)

"Team leaders have pushed off everything to the workers." (Hourly)

"The role of the supervisor is no longer to supervise—workers now do that. Our role now is to do all the work others used to do that are no longer here." (Supervisor(s)

"We are scared—less people, less maintenance, less qualified people in their jobs, less training, more stress. Things are ready to blow." (Supervisor)

"There are no consequences for any behaviors." (Supervisors, Managers)

"Seventy to eighty percent of the employees want rule enforcement bad." (Supervisor)

"This is not a touchy-feeling business—it is just too dangerous. We must have consequences." (Supervisor)

"Those units that were self-directed before still are. Those that weren't still aren't. " (Supervisor)

"Employees have to tell managers what to do;" (Hourly)

"Supervisors choose to be invisible, and choose not to make a decision." (Hourly)

"The only self-direction we get is taking a leak." (Hourly)

"There are too may yes men from the first line supervisor up—they remain politically correct." (Hourly)

"Managers pay is influenced by not spending money." (Hourly)

"Incentives suck." (Hourly)

"Management has intimidated the supervisors." (Hourly)

"Nobody is policing any more. Housekeeping is terrible, and nobody's accountable." (Hourly)

"No definition of what a team leader is supposed to do." (Supervisor)

"We don't know who makes decisions any more." (Hourly)

"Safety is no longer in the supervisor's hands; it should be." (Hourly)

"Nobody makes decisions—it is management by committees." (Hourly)

"We're on the road to disaster." (Supervisor)

"We're just a number." (Hourly)

"We've never got "Self-Directed Work Teams 101." (Supervisor)

"We don't have self-directed teams, just unsupervised teams'. (Supervisor)

"Now the inmates are running the asylum." (Hourly)

And these comments are all from the same company.

REFERENCES

Likert R. *The Human Organization.* New York: McGraw-Hill, 1967.

Peters T, Waterman RH. *In Search of Excellence: Lessons from America's Best-Run Companies.* New York: Harper and Row, 1982.

Petersen D. *Managing Employee Stress.* New York: Aloray, 1990.

11 Assessing Culture

As suggested in the last chapter, the culture of an organization is best assessed by asking the employees how it really is in the organization. Two distinct approaches to this are through interviews or through perception surveys. Using perception surveys is the quickest, least expensive, and most valid. It provides quick, valid data and shows the strengths and weaknesses of the safety culture. Critical safety issues are rapidly identified and any differences in management and employee views on the effectiveness of company safety programs are clearly demonstrated.

Following is a description of one perception survey, typical of others, but the survey this author is most familiar with.

Perception Survey

The goal of the perception survey is to understand how the company is performing in each of 20 safety categories.

The survey begins with a short set of demographic questions which will be used to organize the graphs and tables that show the results (Figure 11-1). Typically the responder is asked about his/her employee level, general work location, and perhaps trade group. At no point is the employee asked questions that would enable him or her to be identified by the people who are scoring the results. Complete anonymity is critical to the success of the survey.

The second part of the survey consists of 100 questions. The questions are

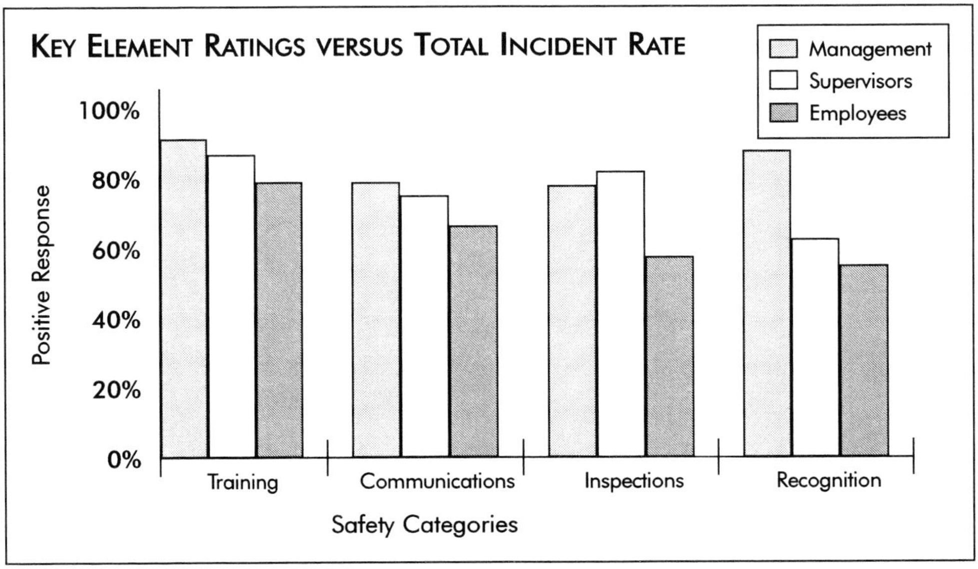

Figure 11-1. Example of perception survey results.

21 Management Systems Areas Measured by the Survey

1. Communications
2. Training
3. Attitudes toward Safety
4. Safety Meetings
5. Support for Safety
6. Inspections
7. Motivational Programs
8. Rules Violation Discipline
9. Safety Regulations
10. Accident Investigation
11. Recognition for Good Safety Performance
12. Involvement of Employees
13. Safety Climate
14. Management Credibility
15. Quality of Supervision
16. Alcohol and Drug Abuse
17. Employment Practices
18. Supervisor Training
19. Goals for Safety Performance
20. Hazards Correction
21. Stress

Figure 11-2.

designed to uncover employee perceptions about the 21 safety categories (Figure 11-2). Questions 1 through 74 focus on the first 20 safety categories and have been statistically, individually validated over 10 years of use in the field. Questions 75 through 100 have recently been added to measure the area of employee stress and statistical validation has not yet been completed. If desired, these questions may be eliminated from the survey without impacting the results in the first 20 safety categories.

All questions are answered with either yes or no. Employees also have the option of leaving the answer blank. Whether a yes answer is considered positive or negative depends upon each individual question.

SCORING THE SURVEY

To score the survey, the number of positive answers for each question are summed and divided by the total number of responses to the question. Questions which are left unanswered by the employee do not impact the final score. This computation provides a percent positive response.

The answers are then correlated with the 21 safety categories. Each question may impact the score of more than one category. A cumulative percent positive response is computed for each category. The percentages for the categories are graphed (see Figure 11-1) to display the results in descending order of positive perception by the line workers. Those categories on the right hand side of the graph are the ones that are perceived by employees as being the least positive and are therefore the most in need of improvement.

Figure 11-1 shows how employee positive responses in Recognition, Inspections, Communications, and Training all suggest weakness in management's accountability—all of which suggests that this organization's culture is not ready to embark on many involvement programs.

Figure 11-3 shows weakness in Supervisory Performance, Management Credibility, Support for Safety, Safety Regulations, Goal Setting, etc. This organization needs to fix current efforts first—and then ask employees to get involved.

It is only when an organization has put in the effort over time to prove it has safety as a core value that management is deemed to be credible. When managers and supervisors are perceived as fulfilling their responsibilities as seen by their daily actions and decisions, then the organization is truly ready for authentic involvement. Thus, it is crucial that the culture be assessed before taking the step toward employee involvement.

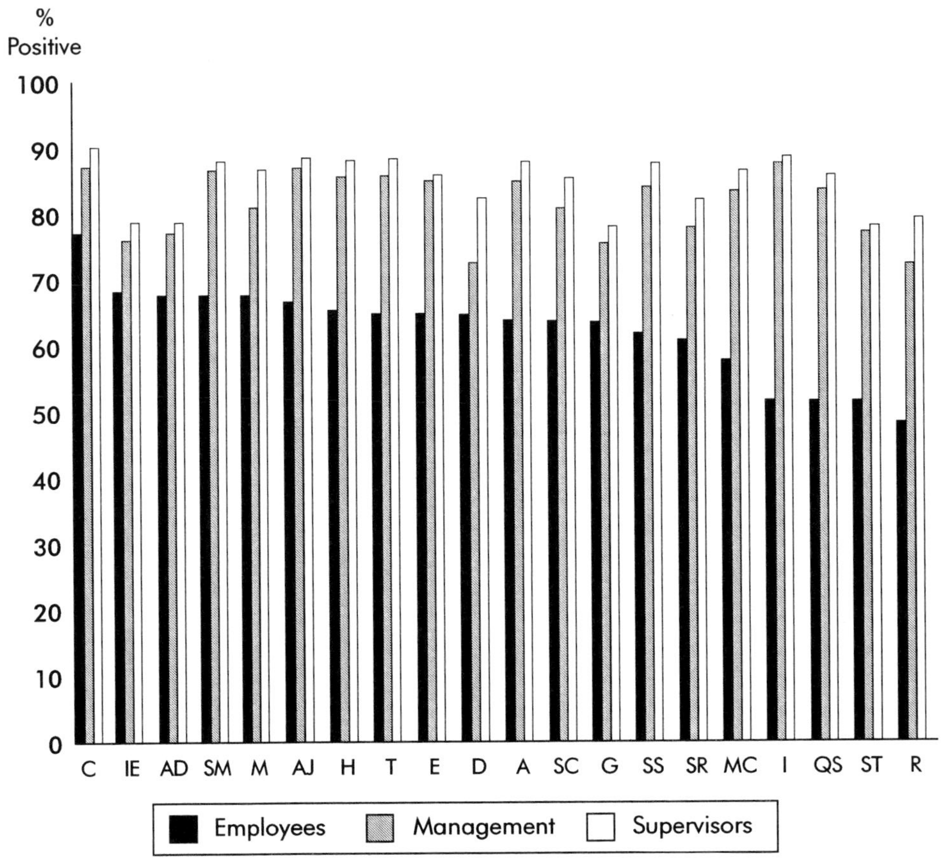

AI Accident Investigation
G Goals for Safety
R Recognition for Performance
QS Quality of Supervision
H Hazard Correction
C Communication
AD Alcohol & Drug Abuse
D Rules Violation—Discipline
I Inspection
SS Support for Safety
SC Safety Climate
T Training
SM Safety Meetings
A Attitudes toward Safety
E Employee Practices
ST Supervisor Training
IE Employee Involvement
M Motivational Programs
SR Safety Regulations
MC Management Credibility

Figure 11-3. Safety survey response.

When the work force believes management is sincere about safety, they invariably will pitch in and do more than their share. So, before embarking into involvement, measure where your culture is, as perceived by the work force. This is crucial.

REFERENCE

Petersen D. *Analyzing Safety System Effectiveness.* New York: J. Wiley, 1996.

12 Behavior-Based Safety

Several times we have mentioned behavior-based safety: in Chapter 1 as the current fad, in Chapter 3 in trying to define what it seems to mean to safety people, in Chapter 4 on what is being marketed, etc. Although this book tries to define involvement in much broader terms than behavior-based safety, perhaps a discussion of behavior-based is necessary, since it seems so pervasive in safety literature, discussions, and especially advertising.

DEFINITIONS

The safety literature, discussions, and advertising use different definitions and descriptions. The following are samples of the variety of definitions:

From Tom Krause of Behavioral Science Technology:

> *The popularization of behavior-based safety management raises the question* What really is behavior-based safety management? *It is important to clearly define the elements that make up behavior-based safety management and how it is distinguished from other approaches. It is important to be clear about just which initiatives are truly behavior-based, and the ones that are merely called that.* [See Figure 12-1.]

Aubrey Daniels of Aubrey Daniels International suggests:

> *The system? It's the same as it's always been. The precise application of the laws that govern human behavior. By identifying what reinforces the*

people in your company and skillfully managing those reinforcers, you can attain the kind of individual performance that adds up to great organizational performance.

(Reprinted with permission from A. Daniels, *Bringing Out the Best in People,* New York: McGraw-Hill, 1994. Copyright by Aubrey Daniels International.)

Scott Geller of Safety Performance Solutions states:

The behavior-based approach starts by identifying observable behaviors targeted for change and the environmental conditions or contingencies that can be manipulated to influence the target behavior(s) in desired directions.

(Reprinted with permission from E.S. Geller, *The Psychology of Safety: How to Improve Behaviors and Attitudes on the Job,* Radnor, PA: Chilton, 1996.)

These quotations are from the gurus of behavioral safety. Not everyone has jumped on board, however. Thomas Smith suggests that behavioral safety (BBS) is doomed to fail:

Systems thinking applied to safety. The way in which employees are interviewed, trained and treated on the job, as well as the design of work methods, materials, machinery and equipment (the system), play a greater role in accidents than unsafe actions. Focusing on the event (the unsafe action) in hopes that it will reveal why the action was taken is doomed to fail. In reality, the systems create the behavior of both managers and workers.

When management realizes that the system itself causes accidents, BBS is no longer useful. In fact, it is detrimental because workers have learned about quality management techniques. On one hand, management freely engages and accepts worker input regarding work systems, while on the other, it applies behaviorist principles to manipulate employee actions. Once workers have learned a better way they won't accept an inferior system.

Therefore, systems thinking must be applied to safety management. As quality theory shows, work is a system—a series of events that must have an aim. When this objective is known everyone works together to achieve it. Customer satisfaction is a key objective. This principle must be applied to safety management.

How a system is managed is a system in itself. BBS fits well into a command and control model, but not within a quality management model. BBS eliminates intrinsic motivation; as a result, it will fail in new quality management systems. In the author's opinion, quality management systems—not

UNDERSTANDING THE SOURCES OF SAFE BEHAVIOR	
The Three Types of Behavior *discovered during observation and two-way feedback*	**The Three Kinds of Barrier Removal** *called for by the three types of behavior*
Non-Enabled Behavior is critical safe behavior that frontline personnel cannot perform because existing systems make the behavior impossible.	Action Planning across levels and areas is the remedy typically called for when a site's existing conditions, equipment, or procedures make a critical safe behavior impossible to perform.
Difficult Behavior is critical safe behavior that frontline personnel can do—but only by going out of their way.	Follow-Up by supervisors and crews is usually required to upgrade a critical behavior from Difficult to Enabled. Removing these barriers to safe performance typically requires specific fix-it items or work orders that do not merit departmental or site-wide action planning, but which frontline personnel cannot authorize or complete by themselves.
Enabled Behavior is identified critical safe behavior that is within the control of the observed frontline personnel.	Observer Feedback is the appropriate form of barrier removal when observed personnel are not using an identified safe behavior that is within their control.

Figure 12-1. The behavior-based safety improvement process.

(Quotation and figure used with permission from T. Krause, The Behavior-Based Safety Process, *2nd ed., New York: VNR, 1997, pp. 9, 53. Copyright © 1997 by John Wiley & Sons, Inc. Reprinted by permission of John Wiley & Sons, Inc.)*

> *BBS—will drive the safety management model that will be used in the 21st century and beyond.*
>
> (Reprinted with permission from T. Smith, "What's Wrong with Behavior-Based Safety?" *Professional Safety,* Sept. 1999.)

Thomas Creswell, founder of the National Safety Management Society, states it a little more forcefully:

> *Back in my days as a general manager, I kicked more behavioral scientists off my facilities than you can shake a stick at. I would not let them come in and screw up my work force. Behavioral psychologists usually have very limited real work experience and can do more damage to an organization*

in a day or week than you can straighten out in 10 years. I have experienced it, I've seen it. I don't feel bad toward these guys for cashing in on their education but I wouldn't give them 15 cents to conduct a class for me on anything. Most of the articles several of the modern day psychologists have written about safety are written to impress themselves. Their 2 dollar words are accurate and what they write is academically correct but they fail to offer a blueprint of how on earth the organization can apply their lofty principles.

(Reprinted with permission from T. Creswell, "Thoughts on Behavior-Based Approaches," *The Communique,* National Safety Management Society, Nov. 1997.)

And Gerald Wagner, from the United Auto Workers, views BBS in this way:

Having persevered through safety slogans, trinkets, posters, games and awards, will the science of injury prevention survive behavior-based safety as well?

Traditional health and safety programs focus on unsafe conditions and utilize a standard protocol termed the Hierarchy of Controls to eliminate/attenuate those unsafe conditions. Conversely, BBS is a four step behavioral model that focuses on the prevention of unsafe acts through a process of behavior observation, personal counseling and reinforcement.

Behavior-based safety is an upstart theory that, as yet, has had minimal impact on traditional injury prevention methodology, but seems to be gaining momentum.

In years past, behavior-based quality management programs which relied on people to never make mistakes were a norm. Today, the paradigm shift is nearly complete. The emphasis is not on prevention of defects by focusing on the process, rather than the persons. As evidenced by the recent popularity of behavior-based safety programs, the paradigm shift in health and safety is not as complete. It is unfortunate that companies are embracing this demonstrably low level of hazard control as their principle injury prevention strategy. Unfortunate, but also understandable. After all, to err is human.

(Reprinted with permission from "The Hierarchy of Controls: An Alternative to Behavior-Based Safety" *Occupational Hazards,* May 1999. Copyright 1999 by Penton Media Inc.)

Obviously there is considerable controversy over behavior-based safety. In a 1999 survey made by *Industrial Safety and Hygiene News,* 46% of those surveyed said they were using behavior-based safety, 54% were not, 49% thought it is here to stay, 51% weren't sure. Apparently the jury is still out.

IS BEHAVIOR-BASED SAFETY INVOLVEMENT?

Our definition of authentic involvement was worker participation where the worker engages in activities that satisfy that person's human needs, and thus are motivational. Every person's needs are different—but most people respond to being treated like mature, adult human beings; being treated as equals, not as subordinates; being able to use their innate intelligence and skills daily, even hourly; being able to achieve; being given responsibility; and being recognized for doing a good job.

Since BBS offers the opportunity for individuals to have involvement only in some things (helping to develop critical behavior inventories, receiving and/or giving training to or with peers, and peer observations) it meets some of the characteristics of authentic involvement, but not all of them.

BBS is clearly not empowerment, nor limited empowerment (freedom to make decisions). It is not self-directed work teams, and is only in part shared decision making. It probably fits best into "providing input so management can make better decisions," which is the lowest form of involvement. Examples of input would be performing job safety analyses, hazard hunts, and ergonomic analysis, or being involved in safety improvement teams.

One other factor to consider is the staying power of BBS. With most of the involvement concepts discussed here, the approaches are integrated into the way the organization works. When the structure of the organization has changed with self-directed work teams or total empowerment, there will be great staying power—it is almost impossible to go back. This is almost true with limited empowerment and shared decision making. It is less true with providing input to management decision making through any of the techniques discussed. These techniques are usually not integrated. They do not become the way the organization is managed. They are more like programs. BBS falls into this category—it can last, but often does not.

REFERENCES

Creswell T. Thoughts on behavior based approaches. *The Communique.* National Safety Management Society, November, 1997, pp. 3-5.

Daniels A. *Bringing Out the Best in People.* New York: McGraw-Hill, 1994.

Geller ES. *The Psychology of Safety: How to Improve Behaviors and Attitudes on the Job.* Radnor, PA: Chilton. 1996.

Industrial Safety and Hygiene News. What do you think of behavioral safety? *IS & IH News.* Sept., 1993.

Krause T. *The Behavior-Based Safety Process: Managing Involvement for an Injury-Free Culture.* 2nd ed. New York: J. Wiley, 1997, pp. 9, 53.

Likert R. *The Human Organization.* New York: McGraw-Hill, 1967.

Peters T, Waterman R. *In Search of Excellence.* New York: Harper and Row, 1982.

Petersen D. *Managing Employee Stress.* New York: Aloray, 1990.

Smith T. What's wrong with behavior based safety? *Professional Safety.* Sept. 1999, pp. 37-40.

Wagner G. The hierarchy of controls: An alternative to behavior-based safety. *Occupational Hazards.* May 1999, pp. 95-97.

13 Future of Authentic Involvement

It is difficult to predict the future of anything, and it is certainly difficult to see where the involvement movement is going. As indicated in the earlier chapters, the involvement concepts have been with us since the 1930s, but they really did not become common until perhaps the 1980s. Deming's popularity helped to push employee involvement. There are some trends in safety today that suggest we will see more and more authentic involvement. For instance:

1. *Acceptance of employee involvement in safety programs.* Most companies are recognizing the importance of employee involvement. Employee-run safety programs are no longer uncommon.
2. *More attention being paid to safety.* Due to the general theme of improving the quality of work life from management books, there are strong efforts toward a people orientation in many (maybe most) organizations. This typically (after time) grows to include safety and health.
3. *Acceptance of accountability concepts.* Most major companies have changed their systems to include clearer role definition, more valid measures for safety performance, and more meaningful reward structures. Most of these changes have been within the last dozen years. This first step toward safety excellence typically is followed by involvement of all employees.

4. *Acceptance of positive reinforcement concepts.* While positive reinforcement is still foreign to safety thinking, at least we have conceptual acceptance to listen and to experiment with it (see Chapter 3, The Rationale for Involvement).
5. *The beginning of a realization that government has not and cannot solve the safety problem.* There is a general acceptance today that after over 30 years of trying, OSHA has made little improvement in the safety record. As we break away from the dictates from government, we begin to utilize newer approaches.
6. *Acceptance of the concept of behavior-based approaches to safety systems.*
7. *Acceptance of the concept that accident statistics are usually a poor measure of safety effectiveness.*
8. *Acceptance of the concept that upstream measures are needed and are a better measure of safety effectiveness, especially when coupled with valid downstream measures.* When we use upstream measures, we can better concentrate on what processes should be in place and how can we measure them.
9. *The beginning of use of behavior sampling as an upstream measure.*

Just as there are positive signs for the future, there are also some signs that barriers remain:

1. *While there is a realization that government does not have the answer to the accident problem; government continues to push harder and probe more and more into areas that many perceive as beyond their charter.*
2. *Some regulations will attempt to require employee participation into companies whether or not they might be ready.*
3. *The advent of criminal liability.* We now find that management has more than financial and moral responsibility for the safety of the worker; in certain situations they now could also end up behind bars. Company executives have been indicted for manslaughter for knowing of hazardous conditions and ignoring them. With this external threat, how many managers will turnover decisions to employees, while retaining sole accountability?
4. *The continued refusal to do anything about the amount of stress that is present in most organizations.* The refusal to recognize stress by management, and to deal with it is truly surprising. As stress continues and increase, people will be less inclined to help.

5. *The continued right-sizing of organizations, in many cases way beyond the optimum, causing additional deterioration of organizational morale and culture, causing people to be less involved in improvement efforts.*
6. *The continued belief in many (perhaps most) organizations they can buy safety excellence through some package they purchase from outside their own people.*
7. *The continued refusal in many (perhaps most) organizations to use the intelligence and skills of their own employees to solve problems and make decisions.*
8. *The terrible waste of management time, effort, and money, through dealing with the latest thrusts of OSHA, rather than on the causes of the losses within their own organization.*

While these plusses and minuses exist, within each organization there are other positives and negatives. Each organization must assess its readiness for involvement. Appendix C suggests some ways to make the assessment.

14 Role of Safety in Achieving Involvement

First, a look at the role of safety generally in an organization, and then some specific thoughts on the safety staff's role in achieving authentic employee involvement.

Safety in the Organization

The goal of management is efficient production—production that maximizes profit. To obtain this goal, management has four basic resources: employees, facilities, equipment, and materials. Management brings many influences to bear on both of these resources. Upon the company's personnel, management applies such influences as training, the selection and placement process, employee health programs, and employee relations. Facilities, equipment, and materials are influenced by maintenance, research, and engineering. These influences and basic resources are brought together through various standard operating procedures. The function of safety staff is to:

- Build safety into these products.
- Continually audit the carrying out of these procedures to ensure that the controls are adequate.
- Help management in the identification of role activities, measurements, and rewards that will lead to excellent results at each level, from the top to the bottom of the organization.
- Effect change when change is indicated.

The safety staff accomplishes these tasks by asking why certain acts and conditions are allowed and asking whether certain known controls exist.

Safety is not a resource; it is not an influence; it is not a procedure; and it certainly is not a program. Rather, safety is a state of mind, an atmosphere that must become an integral part of each and every procedure in the company. This, then, is what we mean by built-in or integrated safety. It is the only brand of safety that is permanently effective.

Since any accident, any unsafe act, or any unsafe condition indicates a system failure, the safety professional must become a systems evaluator. Just stating this, however, does not tell the safety manager what to do, what the actual duties of the staff safety specialist are, or whether he or she has a full- or part-time position. Actually, we cannot. For each of us, the duties that we perform are the duties we have set for ourselves. The job of staff safety specialist is very much a self-defined job. The duties will vary, depending on the size of the organization, the number of locations, the operations themselves, the people above and in line management, the problems presently facing the economy, the other staff people and specialists available, and where the safety specialist fits into the organization. And most importantly today, the duties will vary depending on the culture of the organization, the style of management, the amount of employee involvement, etc.

The American Society of Safety Engineers (ASSE) has spelled out the areas in which the specialist should work to protect people, property, and the environment:

A. *Anticipate, identify and evaluate hazardous conditions and practices.*
B. *Develop hazard control designs, methods, procedures and programs.*
C. *Implement, administer and advise others on hazard controls and hazard control programs.*
D. *Measure, audit and evaluate the effectiveness of hazard controls and hazard control programs.*

(Reprinted with permission from ASSE, *Scope and Functions of the Professional Safety Position,* DesPlaines, IL: ASSE, 1998.)

Beyond these broad guidelines, the safety staff job is self-defined. To help in this definition, the safety manager might refer to Figure 14-1, which briefly outlines the major kinds of activities of the safety staff.

First of all, a system of measurement must be structured so that accountability

Measuring Safety Performance, Using	**Safety Program Development, Including**
• Results Measures • Activity Measures • Audits • Records • Statistics • Inspections • Sampling • Perception Surveys • Interviews	• Orientation • Training • Supervisory Development • Motivation • Gimmicks • Employee Selection Process • Medical Controls • Conditions for Employee Involvement
Being a Technical Resource • In investigation • On standards and regulations * On consumer products * On new equipment	**Being a Systems Analyst**

Figure 14-1. Safety staff's role.

can be fixed and so that rewards can be applied properly to the right people and at the right time in order to reward or reinforce the desired behavior.

Secondly, in safety programming the safety manager must ensure that safety is included in orientation, that safety training is provided where needed, that safety is a part of supervisory development, that the organization's attention is focused on safety, that safety is included in employee selection and in the medical program. When the organization is into employee involvement systems, whether the approach is employee ownership, self-directed teams, or more limited involvement as described in Chapter 4, the safety function must ensure that safety does not fall through the cracks, as often happens.

Thirdly, the safety manager must function as a technical resource: know how to investigate in-depth; know where to get technical data; know the standards; and know how to analyze new products, equipment, and problems.

And fourthly, the safety manager must function as a systems analyst, searching for why things are malfunctioning in the system and whether or not controls are working as intended.

WHAT IS STAFF?

We should probably define more clearly what we mean by safety staff. Staff work consists of assisting an executive in the discharge of executive functions, under the supervision of, on behalf of, and frequently in the name of the executive. The staff does the work that the executive would do with adequate time and the specialized knowledge. This work includes planning, supervising, directing, coordinating, organizing—all, however, under the supervision of the executive. Obviously, since the line has primary safety responsibility, the safety director must be staff. As staff, there is no responsibility for the safety record or the results. Staff responsibilities involve activities which help the line achieve its goals in safety. At first, the safety specialist has no authority over the line, but may have a great deal of influence, which is quite different from authority. How much influence or power will depend on the organization and on the personality of the individual in the line position. The safety specialist in any organization obtains results by using one of two methods: (1) making a recommendation to an executive in the line chain of command and that executive issues an order or (2) obtaining acceptance for suggestions voluntarily from the supervisors without taking the chain-of-command route. More often than not, the safety specialist achieves success by the second route and uses the first route only for rare emergencies.

In high-involvement organizations, the safety personnel become truly leaders to the employees, or employee teams; guiding, coaching, counseling, providing resource and research information, technical guidance, etc.

Most line managers realize that the safety specialist does have stronger influence than that shown on the organization chart. This staff specialist is an expert, has certain status, has management's interest and backing, and, if worst comes to worst, has some influence on management's appraisal of the line manager, and hence on that manager's future. So, although the safety specialist has no authority, there is an element of power. There also are situations where the safety specialist is given degrees of authority over the line. It may be stated in the policy that under certain situations the safety specialist must be consulted and must give approval or can even temporarily step in and assume command (stop the operation). This granting of temporary authority to the staff safety specialist in industry is common and correct. Even so, we must keep in mind that basically the work is accomplished by influencing the line and not by directing it.

Perhaps the key concept in understanding the safety person's role in the organization is the concept of the change agent. The safety professional effects change in the organization in almost any way that is necessary. The safety professional who is a change agent uses his or her organization (or the organization he or she is serving) as a client.

ROLE OF SAFETY IN INVOLVEMENT

As an organization goes down the route of involvement, safety personnel are often threatened and at a loss as to how to control the system. They could develop roles, activities, measures, and rewards for managers and supervisors, but find it difficult to hold a team accountable. They are used to safety rules and OSHA standards, so having teams or individuals making decisions is scary.

Usually organizations go to some form of involvement for profit, production, quality, or cultural improvement purposes, not safety. Safety must accept the new structure and figure out the best way to integrate safety into the new systems.

In some instances, however, organizations have chosen involvement in safety as the lead to involvement in everything else. The Proctor and Gamble story in Chapter 9, Self-Directed Work Teams in Safety, is an example. Dr. Krause and Dr. Geller, mentioned in Chapter 12, Behavior-Based Safety, can provide many examples of the use of behavior-based safety as a lead-in to culture change. In some organizations, safety improvement teams (Chapter 7, Input for Better Decisions) have later led to self-directed work teams.

POWER OF INVOLVEMENT

Examples exist, both good and bad, of the power of employee involvement. In one midwestern energy company, I found teams to be counterproductive because the company established too many teams too fast, resulting in chaos. In interviewing employees at a number of locations, I found that the organization simply wasn't ready for the magnitude of change. Here are some actual quotes from these interviews:

"We never got Self-Directed Work Teams 101." Supervisor

"We don't have self-directed teams, just unsupervised teams." Supervisor

"Now the inmates are running the asylum." Hourly

"Nobody makes decisions—it is management by committees." Hourly

"Going to meetings is more important to management than working." Hourly

"We're on the road to disaster." Supervisor

The above is, no doubt, the worst case scenario. Even in this organization, there were bright spots where involvement worked beautifully. The success is in how the change is managed.

On the bright side, I've encountered what I would consider fantastic results in some organizations. Proctor & Gamble has such results. I've interviewed employees at a Frito-Lay plant in Arizona that had self-directed work teams. The employees were literally excited about what they were doing. At a location of the Union Pacific Railroad on the West Coast, I heard hourly workers make presentations on what their safety improvement teams had achieved, which was excellent—but more important, these workers were feeling achievement and responsibility and being recognized.

Involvement works when we do it right and know it is truly authentic.

REFERENCE

American Society of Safety Engineers. *Scope and Functions of the Professional Safety Position.* DesPlaines, IL: American Society of Safety Engineers, 1998.

V

APPENDICES

Appendix A: Safety Sampling

As indicted in this book, the purpose of employee involvement is really twofold: (1) to tap the knowledge and skills of all employees in making improvements in the safety process and (2) to alter and improve the behavior of all employees. We can measure our process improvements through audits and perception surveys, which tend to be downstream measures. We also need good upstream measures to assess continuous improvement and to assess whether behaviors are actually improving as a result of our efforts. For this, safety sampling is an excellent measure.

Safety Behavior Sampling

One of the methods of fixing accountability, using statistical methods, is safety sampling. Safety sampling measures the effectiveness before the accident by taking a periodic reading of how safely the employees are working.

Like all good accountability systems or measurement tools, safety sampling is also an excellent motivational tool, for each line supervisor wants to be sure the employees are working as safely as possible when the sample is taken. To accomplish this, he or she must carry out some safety activities such as training, supervising, inspecting, and disciplining.

In those organizations that have utilized safety sampling, many report a good improvement in their safety record as a result of the improved interest in safety on the part of the line supervisors.

Safety sampling is based on the quality control principle of random sampling inspection, which is widely used by inspection departments to determine quality of production output without making 100% inspections. For many years, industry has used this inspection technique in which a random sampling of a number of objects is carefully inspected to determine the probable quality of the entire production. The degree of accuracy desired dictates the number of random items selected to be carefully inspected. The greater the number inspected, the greater the accuracy.

Procedure

1. Prepare a code.

The list of unsafe practices is the key to safety sampling and supervisor training. This list contains unsafe acts that occur in your plant. These are the "accidents about to happen."

The list is developed from the accident record of each plant. In addition, possible causes are also listed. The code is then placed on an observation form (see Figure A-1).

2. Take the sample.

With the code attached to a clipboard and with a theater counter, start the sample.

The inspector identifies the department and the supervisor responsible. He or she then proceeds through the area observing every employee who is engaged in some form of activity. The inspector instantaneously records a safe or an unsafe observation of the employee.

Each employee is observed only long enough to make a determination, and once the observation is recorded, it should not be changed. If the observation of the employee indicates safe performance on the job, it is counted on the theater counter. If the employee is observed performing an unsafe practice, a check is made in the column which indicates the type of unsafe practice by the element code number (See Figure A-2).

3. Validate the sample.

The number of observations required to validate is based on a preliminary survey and the degree of desired accuracy. The following data must be recorded on the preliminary survey: total observation and unsafe observations. The

	DEPARTMENT												
Sampling Worksheet Page 1 of 1 Safe Observations Unsafe Acts	DC & Service	Maint. Power	Tool Room	Foundry & Shipping	Rotor	Shaft	Punch Press	Body & Frame	Bracket	Small Winding	Large Winding	Small Assembly	Lg. Assem. & Pck.
1 Improper lifting													
2 Carrying heavy load													
3 Incorrect gripping													
4 Lifting without protective wear													
5 Reaching to lift													
6 Lifting and turning													
7 Lifting and bending													
8 Improper grinding													
9 Improper pouring													
10 Swinging tool toward body													
11 Improper eye protection													
12 Improper foot wear													
13 Loose clothing — moving parts													
14 No hair net or cap													
15 Wearing rings													
16 Finger/hands under dies													
17 Operating equip at unsafe speeds													
18 Foot pedal unguarded													
19 Failure to use guard													
20 Guard adjusted improperly													
21 Climbing on machines													
22 Reaching into machines													
23 Standing in front of machine													
24 Leaning on running machine													
25 Not using push sticks (jigs)													
26 Failure to use hand tools													
27 Walking under load													
28 Leaning — suspended load													
29 Improper use of compressed air													
30 Carrying by lead wires													
31 Table too crowded													
32 Hands and fingers between metal boxes													
33 Underground power tools													
34 Grinding on tool rest													
35 Careless alum. splash													
36 One bracket in shaft piling													
37 Feet under carts or loads													
38 Pushing cars improperly													
39 Pulling carts improperly													
40 Hands or feet outside lift truck													
41 Loose material under foot													
42 Improper piling of material													
43 Unsafe loading of trucks													
44 Unsafe loading of skids													
45 Unsafe loading of racks													
46 Unsafe loading of conveyers													
47 Using defective equipment													
48 Using defective tools													
49 Evidence of horseplay													
50 Running in area													
51 Repair moving machines													
52 No lock-out on machine													
TOTAL UNSAFE ACTS													
Additional Unsafe Acts													
53													
54													
55													
56													
57													
58													

Figure A-1. Safety sampling worksheet.

SAFE PRACTICE SAMPLING REPORT

Plant ________ Period covered ______________

Department Supervisor	Unsafe practice code number														Observations		%
	1	2	7	9	11	17	26	34	36	59					Total	Unsafe	Unsafe
E. Jones — supt.															1,094	39	3.40
Smith — gen. f.															246	9	3.50
Jolas			1					1	1						90	3	3.20
Johnson		3			1			1	1						156	6	3.80
G. McArthur															226	11	4.60
Mantle				1		1		1	1						101	4	3.80
Williams				1	1										53	2	3.60
Nedstrom					1			1	1	2					72	5	6.50
Mack															284	13	4.40
Peters		1			3	1									96	5	5.00
Sadelri							3	1	1						73	5	6.40
Altert	1		1												64	2	3.00
Anderson	1														51	1	1.90

Figure A-2. Safe practice sampling report.

percentage of unsafe observations is then calculated. Using this percentage (P) and the desired accuracy, which we will determine as a ± 10%, we can calculate the number of observations (N) required by using the following formula:

$$N = \frac{4\ (1-P)}{Y^2\ (P)}$$

Where N = total number of observations required

P = percentage of unsafe observations

Y = desired accuracy

For example, if the preliminary survey results produced 126 total observations and 32 unsafe observations, the percentage of unsafe to total observations would be 32 divided by 126, which is 0.254 or 25%.

$$N = \frac{4\ (1-P)}{Y^2\ (P)}$$

$$= \frac{4\ (1 - 0.25)}{(0.10)^2\ (0.25)}$$

$$= \frac{3}{0.0025}$$

$$= 1{,}200 \text{ observations required}$$

To give effective results, this study must have a minimum of 1,200 observations.

4. Report to management

The results can be presented in many different forms. However, the report should include the following:

1. Total percentage of unsafe activity by department and by shift.
2. Percentage of unsafe activity by supervisor, general foreman, or superintendent.
3. Number and type of unsafe practices observed.
4. Types and number of unsafe observations which are supervisory responsibilities.

Appendix B: Measurement

To hold someone accountable, we must know whether he or she is performing well, so we must measure that person's performance. Without measurement, accountability becomes an empty and meaningless concept.

Thus, starting with what we propose as the single most important factor in getting good line safety performance (accountability), we find that we are really talking about ways to measure the line manager better. And measurement has been our downfall in safety for years.

For the line manager, "to measure is to motivate." Although this statement might have sounded a little ridiculous 20 years ago, I believe that it expresses a profound truth, at least in terms of the safety performance of line organizations. Managers react to the measures used by the boss; they perceive a task to be important only when the boss thinks it is worth measuring.

Having perceived the importance of measurement in obtaining good safety performance, we then hit our biggest snag: What shall we measure? Should we measure our failures as demonstrated by accidents that have occurred in the past? If this is in fact a good measure, as has historically been believed (for that is what we usually measure), then what level of failure should be measured? We can measure the level of failure we call "fatalities." Fatalities are used to measure our national highway traffic safety endeavors. Is the measure of fatalities, then, a good measure? Fatalities could be a good measure if we are assessing the national traffic safety picture, but it would be a little ridiculous in the case of a supervisor of ten factory workers. Such a supervisor might well do absolutely nothing to promote safety and still never experience a fatality in his or her department. Obviously, measuring fatalities would make little sense in this case.

While this may sound a little ridiculous, it accurately describes what is going on in many safety programs today. A supervisor can do absolutely nothing related to safety for a year and attain a zero frequency rate with a small bit of luck. By rewarding such a supervisor, we are actually reinforcing nonperformance in safety.

If fatalities (or frequency rates) are a poor measure of supervisory performance, what is a good measure? Or, more important, what is wrong with fatalities as a measure? Perhaps measuring our failures is not the best approach to use in judging safety performance. After all, this is not the way we measure people in other aspects of their jobs. We do not, for instance, measure line managers by the number of parts the people in their departments failed to make yesterday. And we do not measure the worth of sales people by the number of sales they did not make. Rather, in cases like these we decide what performances we want and then we measure to see whether we are getting them.

What would be a good measure of supervisory safety performance? More important, what set of criteria can we develop for measuring supervisory safety performance? Or the safety performance of the corporation? Or our national traffic safety performance? Or anything else related to safety?

Even a brief look at the problem of measurement shows us that we need different measures for different levels in an organization, for different functions, and perhaps even for different managers. What is a good measure for one supervisor of ten people may not be a good measure for another, much less for a plant superintendent or the general manager of seven plants and 10,000 people. What might be a good measure for the supervisor of a foundry cleaning room may be inappropriate for use in judging the effectiveness of OSHA.

Available Measures

Two categories of tools are available: (1) activity (or performance) measures and (2) results measures. As a general rule, in selecting measuring devices, use only activity measures at the lower managerial levels, primarily activity measures (with some results measures) at the middle-upper management levels, and reserve the pure results measures for the executive level. With rare exception (like safety sampling), this rule of thumb gives us statistical validity and is extremely important.

As indicated, we can use either activity measures or results measures to determine performance, and we can use them at the supervisory, the managerial, or the system-wide levels, provided we use caution in measurement selection. At the supervisory level, the activity measures are most appropriate, whether it be the number of inspections, number of people trained, or number of observations made. These activity measures are equally appropriate at the managerial levels, and can even be used at the system-wide levels (through audits or questionnaires). All activity measures are extremely well suited to management by objectives (MBO) or supervision by objectives (SBO) approaches (Were the objectives reached?). Using these measures in SBO assumes that the original objectives were well written (realistic, rifled, measurable, and under the direct control of the objective setter).

Results measures can also be used at all levels, as long as extreme care is used at the lower levels. The traditional safety measures, such as frequency rate or severity rate, cannot be used at the lower levels except over long periods of time and then probably only as a quality check.

MEASURES AT THE SUPERVISORY LEVEL

Performance Measures

We start our examination of supervisory measurement by looking at what the supervisor does to get results and by determining whether the supervisor actually does them.

Performance measures have certain distinct advantages:

- They are flexible and take into account individual supervisory styles. We do not have to use the same measure for all supervisors. We can allow each supervisor to select performance and levels of performance and then measure their performance and levels. Performance measures are excellent for use in MBO approaches.
- They give swift feedback, since most require supervisors to report their level of performance to the boss. (They are also often self-monitoring.)
- They all measure the presence, rather than the absence, of safety.
- They are usually simple and thus administratively feasible.
- They are clearly the most valid of all measures.

MEASURES AT UPPER MANAGEMENT LEVELS

Performance Measures

While we begin to emphasize results measures more at the middle management level, some performance measures should be retained here also. Performance measures at this level are simple measures of what middle managers do, whether they perform the tasks that are necessary for them to perform. Thus, everything we have said about supervisory performance measures applies also at the middle management level.

However, supervisory performance and middle management performance should be quite different, necessitating some changes in the measures used at this level. Normally, we want the middle managers to get their subordinates (supervisors) to do something in safety. Thus, we can measure them to see whether they meet with their supervisors, check on these supervisors, or monitor the quality of the supervisors' work.

Audits

A measure of performance (and results) that enters the scene at the upper and middle management levels (management of a location, for instance) is the audit of safety performance.

Failure Measures

Failure measures tend to be generated from our injury-record system.

In order for injury records to be used for measuring the results of the line manager's safety performance, they should be set up so that:

- They are broken down by unit.
- They give some insight into the nature and causes of accidents.
- They are expressed eventually in terms of dollars by unit.
- They conform to any legal and insurance requirements.

Beyond these broad outlines, each company can devise any system that seems right for it. The dollar criteria is included because of the belief that a dollar measuring stick is much more meaningful to line personnel than any safety specialist's measuring stick (such as our frequency and severity rates).

Before-the-Fact Measures

Before-the-fact measures measure the results of a supervisor's action before an accident occurs. For instance, a periodic inspection is made of a supervisor's work area to measure how well he or she is maintaining physical conditions. This is a measure of whether things are wrong and, if so, how many things are wrong. We can also measure how well a supervisor gets through to the people in the department by measuring the people's work behavior (safety sampling).

One of the more recent and best measurement tools is the perception survey.

RESULTS MEASURES AT MIDDLE AND UPPER MANAGEMENT LEVELS

At the middle management level, we concentrate on results measures. Here again we can work with before-the-fact measures or with failure measures. The validity and reliability of our failure measures may depend upon the size of the unit we are working with. Before-the-fact results measures used to judge a middle manager's performance would include safety sampling and inspections by staff safety. Failure measures can be used a bit more extensively at this level. Here our traditional indicators become more useful.

Appendix C: Assessing Culture

Since culture is defined as "the way it is around here," then the way to assess it is to go to the people who know—the hourly employees. Audits assess our systems (Are they in place?); they do not assess their effectiveness. Our statistics offer little insight unless we are generating a massive number of injuries.

Thus, to assess culture, we must ask the people who know, the hourly employees, and we must ask them in a manner that gives us results in a form that directs us to improvement. We have two basic approaches: to interview or to use a perception survey, as suggested in Chapter 11.

This appendix offers some actual examples of culture assessments from these two sources.

CULTURE ASSESSMENTS THROUGH INTERVIEWS

Example 1. Results of Interviews from a Large Utility

This study interviewed 225 people at 20 different locations. The people interviewed included the following, with the results listed below:

10	Executives	4	Union Heads	1	Legal
15	Plant/Area Manager	99	Hourly	41	Meetings Staff
39	First-Line Supervisors	12	Safety Staff		
3	Human Resources	1	Medical		

1. Safety is not perceived to be a company value by anyone interviewed. It is in the perceived list of priorities, with 89% of the hourly ranking it behind production, and cost containment.

2. Top Management. People stating upper management as being credible in safety is low ranging from 0% at the hourly level to a high of 30% at the highest levels interviewed.

3. Leadership. Supervisors (Team Leaders) are not performing in safety.

- *They are spread too thin.*
- *Some don't think that they should anymore.*
- *They are measured under the performance management system by accident statistics (luck) and by special projects; they are not measured meaningfully or rewarded for supervising or leading.*
- *They are performing tasks that need to be done by support people.*
- *Most of their time is spent in committees or meetings.*
- *In interviews with hourly people, 89% stated that they were still having safety meetings (100% of supervisors and mid managers said they were happening). Zero percent of hourly people said there was any other type of supervisory involvement in safety. The consensus of most hourly people interviewed was that 75% of the leaders are invisible in the field or plant.*

4. Teams. To a large degree unsupervised, teams, with rare exception, have not filled the leadership void. Some who have attempted self-direction have been overridden by bosses or the decisions made. Some simply work unsupervised, but now believe they can choose to work in any manner they want, including taking risks to speed the work.

5. Employee Ownership. The attempt to get hourly employee ownership through re-engineering has been interpreted in many different ways. Some have accepted it, many have not. Some areas have always had it. Some were ready (mature enough). The more experienced handle it the best—but many of them are no longer on the job, creating a void.

The hourly employees have all heard from management that safety is important; that they should always take all the time needed to do the job safely; but throughout the entire system there is an overriding belief that there is so much work to be done that they must rush and take chances and that is what management really wants.

6. Management of Safety. Safety at this time is personality dependent, not system driven. If a boss or team believes in safety, it gets accomplished. There are many good managers, supervisors, and employees, thus bright spots. But these are the exception, not the rule.

7. To a large degree, with some exceptions, there is management by committee, not management by accountability.

8. Committees, Safety and Other.

In most interviews, particularly at lower levels, one of the problems spelled out was too many committees, allowing management to abdicate decision making, and taking an excessive amount of time off the job, further stressing a downsized work force. Whereas supervision does not have time to supervise, 100% of those interviewed were on at least one committee. Hourly employees also serve on many committees. This creates a problem where uninvolved workers must do the work of the involved, a major source of irritation.

This organization started an employee involvement effort when they clearly were not ready. The results were chaotic. They needed to fix the management problems first.

Example 2. Results of Interviews from a Large Railroad

This study interviewed 341 employees: 118 in management, 220 agreement employees, and all their top executives. Eight locations were visited, three regional offices plus the corporate office. The results were as follows:

Safety Leadership:

- *Roles are not clearly spelled out at most levels. Some activities are defined in safety action plans, but there is no real quality check on those activities.*

- *No champion for safety in executive management.*
- *Executives at the system level are not visible. Executives at the regional level are often too visible (due to conference call frequency and length, intimidation, etc.).*

Safety Performance:

- *The primary measure is still the accident rate, which does not drive improvement in the process. There is no correlation study made to determine program value.*
- *Performance targets are defined by safety action plans and job agreements. There is more emphasis on quantity than quality of performance. Feedback from employees is either not there or not listened to.*
- *Discipline guidelines are set. Accident investigation is weak in that root causes in the management system are ignored.*
- *Safety performance is not tied to either compensation or promotability in any meaningful way.*
- *Communication. One on ones are performed, but not often enough. Town hall meetings are held.*
- *Training. Training at all levels is weak.*

These items are addressed at the local level through specific programs or "islands of safety," such as:

- *Personal Injury Reports*
- *Efficiency Tests*
- *Team Tests*
- *Discipline Policy*
- *Buddy Program*
- *1-5 Year Program*
- *One on Ones*
- *Operation Lifesaver*
- *Safety Contacts*
- *Safety Action Plans*
- *Job Agreements*
- *Safety Meetings*
- *Safety Hotline*
- *Recognition Hotline*
- *Spot Awards*

These programs are managed primarily by conference calls, where data is collected or checked, questions answered, reinforcements provided, CYA takes place, etc.

The mistakes in this process include the following:

- *Roles are not clearly defined for each position*
- *Poor measures are used in performance review*
- *Little training takes place in most locations*
- *Little time is available for field work*
- *Lots of pressure and stress*
- *Lots of hours per day at work*
- *Little improvement in relationships*

The Safety Manual is a quality system procedure; however, in actual practice there is little evidence of the quality process on safety. The principles of quality management are largely ignored in the safety process. Improvement teams are rare, particularly at the agreement level. Root cause analyses are also rare, and there is little evidence of the use of analytical tools to solve problems. About the only use in safety of the tools is the use of paretos on accident data, and even here there is no use of control charts.

There is a need for upstream, nonaccident measures of process improvement at all levels, and a need for regular perception surveys (or interviews) downstream. In short, there is no real measure of how well the safety process is working.

The Safety Manual lays out the safety action plan contents, allowing little flexibility for local impact and innovation; thus safety action plans tend to look alike. The quality of the written plans, however, has little correlation to the accident records of the unit.

This organization has considerable work to do on the basics of safety. They were tightly involved with TQM concepts and procedures in every aspect of management, except safety. Here the employees were ready for involvement but management was not.

CULTURE ASSESSMENTS THROUGH PERCEPTION SURVEYS

Example 1. Results of Survey of a Large Utility

The results of the perception survey (employee only level, percent positive by category) are as follows:

Discipline	*60%*
Recognition	*65%*
Inspections	*72%*
Alcohol and Drugs	*75%*
Supervisor Training	*76%*
Awareness Programs	*77%*
Operating Procedures	*78%*
Involvement	*79%*
Attitude	*80%*
Employee Training	*81%*
Quality of Supervision	*81%*
Support	*82%*
Goals	*82%*
New Employee Orientation	*83%*
Safety Climate	*84%*
Communication	*84%*
Management Credibility	*87%*
Safety Contacts	*87%*
Hazard Correction	*90%*
Accident Investigation	*92%*
Overall	*80%*

Overall, the scores were quite high, with only two categories showing any concern. The results suggest for the most part very good control. The categories of discipline, recognition, inspections, supervisor training, quality of supervision, and safety contacts are indicators of supervisory performance. These six average out to 73%. Still quite good, but just over the 70% point of concern.

The categories of support, management credibility, operating procedures, and goals tend to indicate the perception of management generally, and these four average out to 82% positive, which is very good.

When we compare these results to a recent study made of all companies who have taken this perception survey, we find the following results:

		Other Companies
Discipline	*60%*	*58%*
Recognition	*65%*	*57%*
Inspections	*72%*	*60%*
Alcohol and Drugs	*75%*	*64%*
Supervisor Training	*76%*	*60%*
Awareness Programs	*77%*	*73%*
Operating Procedures	*76%*	*60%*
Involvement	*79%*	*66%*
Attitude	*80%*	*70%*
Employee Training	*81%*	*65%*
Quality of Supervision	*81%*	*65%*
Support	*82%*	*70%*
Goals	*82%*	*71%*
New Employee Orientation	*83%*	*73%*
Safety Climate	*84%*	*72%*
Communication	*84%*	*76%*
Management Credibility	*87%*	*70%*
Safety Contacts	*87%*	*73%*
Hazard Correction	*90%*	*73%*

Accident Investigation	*92%*	*78%*
Overall	*80%*	*68%*

As you can see, the company compares very favorably in every category.

The results of this perception survey suggest that in most aspects of the safety system controls are in place. The only areas suggesting some attention currently are the discipline process and the amount of recognition employees get routinely. In both cases, you might want to look at the individual question scores in each category for further insight. In both categories, both the process and the implementation by supervisors might be examined.

Overall, employee scores are quite high compared to other companies (18% higher overall). This is excellent, suggesting the current direction is working.

This company clearly is now ready for more employee involvement programs. There is good trust between management and the work force. The difference in perception between employees and managers was small. They will be successful with any form of involvement.

Example 2. Results of Survey of Furniture Manufacturer

The results of the perception survey were as follows:

For the entire organization, the worst categories in positive employee response are:

Inspections	*53%*
Safety Contacts	*53%*
Goals	*55%*
Recognition	*58%*
Quality of Supervision	*59%*
Involvement	*60%*
Alcohol and Drugs	*60%*
Stress	*63%*
Awareness	*63%*
Employee Training	*63%*
Supervisor Training	*64%*

Discipline	*64%*
Communication	*65%*
New Employee Orientation	*65%*
Operating Procedures	*66%*
Support	*66%*
Accident Investigation	*68%*
Hazard Correction	*68%*
Attitude	*71%*
Management Credibility	*72%*
Climate	*73%*
Overall	*63%*

Systemic Problems

There is a general and pervasive feeling throughout the organization that nobody (management, supervision, or the workers) supports the safety effort. Almost one-third feel and express this lack of support. Safety clearly has not been and is not either a key value for the organization or a priority. The sense is that it has been ignored.

Discipline is stated as a problem—lacking and inconsistent. Employee training has been lacking (63% positive), and according to 40% of the work force, alcohol and drugs are not being effectively handled.

Supervisory performance in safety is ineffective according to 40% of the employees (also shown by inspections, safety contacts, recognition, etc.).

Things needed for excellent results in safety are relatively simple. It requires these things be in place:

- *A firm and solid management commitment from the top of the organization, visibly demonstrated through action and decisions.*
- *An accountability system that for each supervisor, manager, and executive defines clearly the expectations (tasks and activities that must be done each day); ensures each person knows how to perform the required tasks through training; measures whether or not the required tasks have been completed; and rewards for completion. That is:*
  - *Roles must be spelled out for each.*
  - *Training must be given where needed.*

- *Activity measures must be used instead of accident statistics.*
- *Safety performance must be included in performance appraisals and daily reporting.*

- *Systems must be in place that ask for and get the involvement of as many hourly employees as possible.*

According to this perception survey, the employees are stating that none of these essentials is in place. The first steps should be to concentrate on the following three areas:

- *Management commitment*
- *The accountability system*
- *The involvement of the work force*

Following this, there are a number of essential areas in need of examination and improvement:

- *The system of discipline (64%)*
- *Supervisor training (64%)*
- *Employee training (63%)*
- *The alcohol and drug program (60%)*
- *Reducing stress levels (63%)*
- *Inspection procedures and hazard correction (53%)*

The differences in perception between levels also suggests the need to adopt some systems to ensure that managers and supervisors do interact with employees to know the reality of what is working and what is not. A 45% difference in perception between managers and hourly employees on inspection; a 36% difference between supervisors and employees on quality of supervision; a 35% and 31% difference in accident investigation; all suggest the need for managers and supervisors to become more cognizant of what the employees are seeing and feeling.

This company has a long way to go in all of their safety efforts. They are probably a long way from being able to get involvement. They might, however, build some involvement in right at the start of a whole new safety thrust by asking hourly employees to help in structuring their safety system.

In all four examples these were baseline surveys. In all four, improvements resulted in the system where indicated in the survey.

Typical Cultures

One advantage of using perception surveys is that an organization can compare its safety efforts to those of other organizations. A recent study of 56 companies (covering 1.657 million employees) allows for some comparisons and offers a picture of safety effectiveness (and culture) in typical United States companies.

The results (employee only positive response) were as follows:

- Only two categories scored below 60% positive:

Recognition	56.9%
Discipline	58.4%

- In seven categories the score was below 70% positive:

Inspections	60.0%
Supervisory Training	60.3%
Alcohol and Drugs	63.7%
Employee Training	64.6%
Quality of Supervision	65.4%
Employee Involvement	66.4%
Operating Procedures	67.6%

- Scores were borderline in the following categories:

Attitudes toward Safety	70.0%
Support for Safety	70.0%
Management Credibility	70.0%
Goal Setting	70.7%

- Overall, we are relatively good in these categories:

Accident Investigation	78.0%
Communication	75.7%
Motivational Programs	73.4%
New Employee Orientation	72.7%
Safety Contacts	72.7%
Climate	72.3 %

This, perhaps, is a good picture of the state of the art in safety. Employee involvement is relatively low, but in many companies management performance is even lower, suggesting that in those companies some confidence and trust needs to be built, and then involvement can be improved.

List of Figures

LIST OF FIGURES

Index